EASTERN EUROPEAN TECHNOLOGICAL BUSINESS OPPORTUNITIES:

AN EXECUTIVE REPORT

by
John J. Langwick

Published by
THE FAIRMONT PRESS, INC.
700 Indian Trail
Lilburn, GA 30247

Library of Congress Cataloging-in-Publication Data

Langwick, John J.
Eastern european technological business opportunities: an executive report / John J. Langwick
p. cm.
1. Europe, Eastern--Economic conditions--1989- 2. Europe, Eastern--Commerce--Directories. I. Title.
ISBN 0-88173-170-6

HC244.L33 1992 330.947--dc20 92-39409
CIP

Eastern European Technological Business Opportunities: An Executive Report by John J. Langwick.

Published by The Fairmont Press, Inc.
700 Indian Trail
Lilburn, GA 30247

Printed in the United States of America

10 9 8 7 6 5 4 3 2 1

ISBN 0-88173-170-6 FP

ISBN 0-13-605890-6 PH

While every effort is made to provide dependable information, the publisher, authors, and editors cannot be held responsible for any errors or omissions.

Distributed by PTR Prentice-Hall, Inc.
A Simon & Schuster Company
Englewood Cliffs, NJ 07632

Prentice-Hall International (UK) Limited, London
Prentice-Hall of Australia Pty. Limited, Sydney
Prentice-Hall Canada Inc., Toronto
Prentice-Hall Hispanoamericana, S.A., Mexico
Prentice-Hall of India Private Limited, New Delhi
Prentice-Hall of Japan, Inc., Tokyo
Simon & Schuster Asia Pte. Ltd., Singapore
Editora Prentice-Hall do Brasil, Ltda., Rio de Janeiro

EASTERN EUROPEAN TECHNOLOGICAL BUSINESS OPPORTUNITIES:

An Executive Report

TABLE OF CONTENTS

CHAPTER 1

EXECUTIVE SUMMARY

1.1 Profile of Eastern Europe

Eastern Europe currently holds some of the most promising and exciting markets anywhere in our ever-shrinking world. The countries that form this region require the complete spectrum of products, goods, services and assistance available in order to rebuild their economies, re-establish failing infrastructure and upgrade their sagging standards of living.

Decades of centralized government have effectively stifled the human desire to excel, improve its position and standard of living. Although the governments of these centralized economies were running the countries on the surface, they were unable to dominate the pride and emotion that drove the people to survive. Collectively, the countries previously known as Eastern Europe are composed of people who are proud of their long and deeply rooted history. The suppression of these popular and nationalist feelings was never completely successful. The people that made up the countries of Eastern Europe were able to establish a second, or invisible, economy through which they kept their families, their thoughts and their spirits alive. This indomitable spirit is the key to their future and the rebuilding of the ruins left behind by the failure of the centralized governments and their planners.

The development of the manufacturing, industrial and transportation infrastructure occurred primarily during the 1950's and 1960's. This was the major push by the Soviet dominated governments of the region to establish what they envisioned to become the world's major industrial power. The idea and the dream never translated into the dominant and interactive single economy that was visualized by the central planners. Their failure has been evidenced by the collapse of the centrally planned Eastern European governments and their replacement with very nationalistic popular governments.

The new governments have inherited conditions and problems which have taken five decades to develop. Their manufacturing technology and facilities are outdated and very inefficient by today's standards. Energy production, distribution and utilization have not kept pace with the last forty years of development and efficiency innovation. The results are goods that are not needed, quality that prevents distribution on international markets, extremely high energy costs, and environmental problems that are recognized as the worst in the world. The air and water pollution problems, for example, have made some regions nearly uninhabitable.

The business and industrial sectors demonstrating the greatest needs and opportunities are related to energy and environment.

1.2 Energy

Energy may be the single most important sector for development in all of the Eastern European countries. Improvement in this single sector will probably show greater results in reduced costs of production, reduced environmental pollution, increased competitiveness and improved standard of living than improvements in any other sector. Eastern Europe represents an enormous new market for energy-efficient products of all kinds. Potential products include diagnostic and monitoring equipment, insulation, energy management control systems, high efficiency burners, lighting equipment and controls -- in fact, virtually every energy-efficient product currently manufactured.[1]

The technology used throughout their energy production, distribution and utilization systems is outmoded and inefficient based on today's most common technology. As deregulation dramatically increases fuel costs throughout the region, including the new "Commonwealth of Independent States," it is expected that energy costs will rise up by 1,000 percent from former subsidized prices. This change will quickly move energy costs out of reach of the consumer and industry, reducing any competitiveness and the general standard of living. It is expected that demand for energy use, specifically electric energy, will decline from 20 percent to 40 percent. The decreased demand will result declining of industrial production. This decline should be expected as the system reacts to the changing requirements of a market-based economy. As the new market-based economy develops and grows, this reduced demand will stabilize and then grow. The new demand will be realized when the need for cost control, production management, quality control and marketing becomes established parts of the new system. It is estimated that it could take 10 to 15 years for these changes to become part of the system. The reduced demand will be what provides them with the time to explore and incorporate new technology for more efficient energy production, distribution and utilization. The Alliance to Save Energy estimates the market for energy-efficient products in Eastern Europe to be $20 billion.[2] Additional information concerning the energy and costs can be found in Appendix T.

1.2.1 Production

Electrical production in the countries is adequate for current requirements. Reduced demand from industry has created excess capacity that will provide a comfortable margin in generating capacity for the next 10 years. This should provide for the temporary shutdown of selected facilities to update the technology and efficiency of electrical generation. The technology still employed in the majority of the power plants is now twenty-five to forty years old. The deregulation of fuel price will be increasing fuel costs by three hundred to four hundred percent. Monitoring and control systems have to be brought up to date to allow continued operation under the new fuel price structure. Burners and burner technology, fuel preparation and preprocessing, and fuel ignition systems have to be upgraded with more current and efficient systems. The reduced fuel usage will be a major cost savings and will reduce environmental problems associated with poor fuel utilization.

1.2.1.1 Coal-Fired Plants

Coal-fired power plants produce the majority of available electrical power. The lignite or brown coal that is common in the region is a major contributor to the severe air pollution problems. Scrubber technology for the stacks and new technology to process cleaner burning fuels from the lignite will have to be employed. See Appendix U for more complete information on the number of plants and the total kilowatt generation capacity of these coal-fired plants in each of the targeted countries.

1.2.1.2 Gas- and Oil-Fired Plants

Gas- and oil-fired plants have been brought on-line in each country. All the fuel for these power plants must be imported, mostly from the CIS. Price deregulation and CIS domestic needs make the supply of gas and oil very expensive and possibly unreliable. Dependence on the facilities and the expenditure to upgrade this will not be of the highest priority.

1.2.1.3 Nuclear Power

Nuclear power is used to produce a major portion of the electrical power in some of these East European countries. Nuclear energy is being produced in plants of Soviet design and of the Chernobyl graphite type. Political action in each of the countries is trying to shut down these existing plants. In many cases it is expected that the political action will cause the further use of nuclear energy to be banned. Should these plants be shut down, their demolition and disposable will create opportunities for advanced technology and systems. The nuclear power plants which remain on-line, will require major work and refit to incorporate safety systems, controls, and monitoring systems. The minimum effort to upgrade the existing plants will require a huge commitment by each of the respective new governments. Yet the continued operation of these plants will provide enormous opportunities for state-of-the-art systems and controls for the safe and efficient operation of the plants.

1.2.2 Distribution

Distribution is accomplished through grids established independently on a regional basis in each of the countries. The grids are then tied together for mutual support of the whole region and the sale of excess power as demand may exist in other distribution grids. Most of the technology used in the development and operation of the grids is obsolete and inefficient. There are a variety of transmission line sizes and capacities used throughout the region. This causes compatibility problems in the linkage of grids. There are plans for standardization, but financing and technology have not been available to allow the systems to be upgraded and standardized. This is a major opportunity for firms that elect to get involved early in the process. Additional information is in Appendix U.

1.2.3 Utilization

Utilization of energy, specifically electrical energy, is reminiscent of U.S. industry in the 1940's. Industry in the region has done very little to improve the efficiency of energy utilization, since it had been abundant and extremely cheap. Deregulation of fuel costs will have a dramatic effect on costs. The increased cost will put pressure on all sectors to change the level of technology being employed for the use and control of energy application. One example of extreme waste of energy is in the use fuel and plant for central heating systems for residences, offices and factories. Very few buildings have active thermostatic temperature control. Buildings are heated and the temperature is controlled by opening windows to disperse excess heat. There are estimates that industrial use of energy can be reduced by over 60 percent with no loss of production. Estimates for possible savings in residential and office heating costs are even greater. With energy prices rising four hundred to six hundred percent within two years of price deregulation, the efficient use of energy will be critical to the survival of their industries. The same or larger increases in efficiency can be expected in energy utilization in other sectors.

1.3 Environment

The environment has deteriorated throughout Eastern Europe. Some of the worst cases of pollution of air, water and land are now being documented in these areas. Most of the world was previously unaware of the severity of the environmental disaster taking place under the protective cloak of centrally controlled governments of Eastern Europe. Now that the door has been opened, even those who have traveled extensively in these countries are shocked at the degree of environmental damage that has taken place. The air, water and soil pollution in Czechoslovakia is said to be the worst in the world. Uncontrolled emissions have denuded landscapes, dramatically increased rates on respiratory diseases, diminished wildlife and reduced agricultural production. According the Czech Green Party, the pollution is the result of industrialization which has run wild and governments which had no responsibility to the people. The Green Party said that statistics speak of reduced life expectancy; birth defects afflict as many as 10 percent of infants in Northern Bohemia.[3]

The disposal of toxic waste is another major but still unmeasured problem. The withdrawal of the Soviet military has left behind more than its share of contamination. In Czechoslovakia, officials stated, "What the Soviets are not taking home is their waste. Tons of it. We still don't know how much damage the Soviets did to their land in the 20 years they occupied Czechoslovakia, but experts estimate that it may take 50 years to clean up the garbage dumps that smolder in the woods, the crumbling buildings infested with vermin and the pools of oil, fuel, and solvents that have leached into groundwater, poisoning wells and introducing heavy metals and carcinogens into the food chain." The deputy environment minister said, "No one knows how much waste remains on the 132 bases the Soviets maintained in Czechoslovakia. The biggest fear is that in some areas there are buried, expired ammunition and chemical weapons with no maps of where they have been left."[4]

Cities like Prague and Czechoslovakia have such severe air pollution that many days it resembles heavy fog overhanging the city.

Tennessee environmentalist, Rick Ellis said, "This is dense petrochemical smog. This is the worst I've seen it anywhere is the world". He adds the forty years of uncontrolled automobile exhaust and heavy industrial pollution have created grave dangers for Czechoslovakia.[5]

1.4 Discussion

U.S. companies with interests in any of these fields will find the opportunities to be staggering. Even now we are only beginning to see what really needs to be done. In many cases, we still have not been able to grasp the enormity of the problems that have until now been classified **TOP SECRET.** Any delay in getting involved early in this process will be very costly to you and your company. A delay now, in the beginning of the process, may even limit your access to the market.

These countries, which have become known as Eastern Europe, are made up of the former COMCON countries of Poland, Czechoslovakia, Hungary, East Germany, Romania and Bulgaria. Many times Yugoslavia is included although it was never a member of COMCON, nor was Yugoslavia held as one of the former Soviet satellite buffer states.

The world was shocked in 1989 when a movement for freedom and independence welled up and spread throughout the region. The economic and political reform which has been sought for decades is now becoming a reality, but it will be difficult. The political reforms, as difficult as they will be, are easier to see and direct than the more basic and urgently need economic reforms.

The economic reforms required in each in these countries will provide considerable business and investment opportunities for firms in the energy, environmental, food and agribusiness fields. These industries are now seen as the basis for re-establishing their economies. The application of state-of-the-art technology and systems will require the knowledge, expertise and equipment from the United States, Western Europe and Asia.

Time is of the essence. The competition to be involved will be intense. U.S. firms that make their move now to get involved will find it to be satisfying and profitable. The economic rebuilding will take years, but it will be conducted by and with those companies who seek opportunity early and establish their credentials and relationships in the market place. U.S. officials say that American firms are too timid. East European firms do not want to be dominated by German firms; they want to diversify their contact and sources. They would like to see contracts, goods and expertise from around the world at work with them in the redevelopment of their countries. More expressly, most East Europeans are outspoken about wanting to have heavy American involvement in the rebuilding and redevelopment of their infrastructure and economies.

Of the countries in the region, three currently are of significant interest. Hungary and Czechoslovakia have a history of trade and a strong desire for international relations, even under the strict Soviet control under which they operated. Poland has been a country where the people lived and operated on two entirely separate levels for many years. The government or official level was the first and most visible level. The government was central in all business transactions and provided nearly all the employment. Buying, selling, production, and manufacturing as well as all goods and services were to be through one of the many government agencies. A second level developed within the Polish economy. This became what could be called a shadow economy. This second level trade and commerce thrived despite the oversight and control of the government. As a result of these experiences, Poland, Hungary and Czechoslovakia are now better prepared to accept the rigors of a free market system. They have a better understanding of the time, energy and capital demands required by a supply and demand system.

Poland, Czechoslovakia and Hungary have stronger economies supporting their ability to effect these changes. The new governments have taken steps to open their markets by reducing trade barriers, providing investment incentives, foreign ownership, repatriation of profits, the establishment of convertible currencies, and the development of international banking relationships. These changes are a strong attraction to firms which had recognized business potential created by redevelopment of infrastructure and the demand for industrial and consumer goods and services. In reality, billions of dollars are in the short term and hundreds of billion dollars are in the next twenty years. This kind of potential is available now for those companies that have the foresight to get involved. Those that take the wait-and-see approach to making contacts and researching the market for their own niche will find it much tougher later. As in any developing market, the original contacts made by those with only slightly greater interest and drive will have established relationships on which to draw as the business and redevelopment projects become reality. As in other areas of the world, the people making the decisions will remember who had the foresight, the drive and the ingenuity to seek them out and offer their services. These will be the same individuals who are asked to fill the contracts to supply the necessary goods and services.

The German Democratic Republic (East Germany), Bulgaria, Romania, and Yugoslavia have not been selected for inclusion in this report. Although these countries also have demands for goods, services and the development of infrastructure, it was felt that each of these countries had unique reasons that made it less attractive as commercial markets at this time. We strongly believe this situation will change in the short term. Even at this writing, decisions are being made in each of these countries that will attract outside investment and the development of viable systems through which their developing private sector business will grow strong. The development of that strong private sector will direct and manage procuring, manufacturing and marketing of goods and services to and from their individual economies.

In the case of the German Democratic Republic (East Germany), it was to be absorbed in the Federal Republic of Germany (West Germany) during the highly publicized reunification of Germany. The result is that the German Democratic Republic does not continue to exist as a country or an independent entity. The German Democratic Republic does continue as part of the reunified Germany under the government and system of the Federal Republic of Germany.

Bulgaria and Romania both have needs that are equal to or greater than the countries which have been selected above. The economic depression which these two countries have experienced during the last several decades has left their primarily agricultural economies without a strong industrial sector. The absence of infrastructural development and the mismanagement of national resources have left these economies in shambles. It may take these countries several years before they are in the position to make the changes and establish the stability required in their political and economic systems to attract outside expertise and investment. At this time we suggest that you watch for opportunities in these economies, but concentrate your efforts on the three countries which have been selected for the report. As you become established in the region, it will put you in a better position to monitor and then enter these additional market countries when the time is right.

Yugoslavia should be an excellent opportunity, but the vicious nature of their current civil wars has destroyed their economy and much of the highly developed infrastructure which had been built during the last several decades. Again, we recommend that you follow developments and watch for the future opportunities which occur in this country.

CHAPTER 2
EASTERN EUROPEAN PROFILE

2.1 Centrally Planned Economies

The centrally planned economies of Eastern Europe and the Soviet Union have now been shown to be inoperative and inefficient. The downfall of these centrally planned economies was their inability to adjust to the ever-changing economic and political requirements of business, industry, their people and the world around them. As a result, central planning has been the inattentive author of massive waste and the now-recognized environmental catastrophe which is becoming more evident throughout the region. The waste has sapped the economy and impoverished the citizens, while the environmental calamity has reduced entire areas to being nearly uninhabitable. Despite these obstacles, the spirit and desire of the people have grown. The people are now demanding the opportunity to invigorate their lives and build an effective resurgence of their economy and standard of living for themselves, the children and the neighbors.

The effort will take decades of concerted effort and hundreds of billions of dollars of investment for redevelopment of the infrastructure, decontamination of the environment and the rebirth of happy, healthy, well-educated populations. The governments of the United States and the other developed and industrialized countries are making concerted efforts to provide the assistance needed to make the massive transformation a reality.

The major support for the effort to reverse the effects of central planning and help re-establish strong and efficient market-based economies is being provided by many countries and international groups. U.S. government agencies and institutions, along with their counterparts in other G-7 countries, are providing direct and indirect funding and guarantees to start the process of economic revitalization in the targeted countries. Additional assistance is being provided by World Bank, area development banks, Enterprise Funds, the IMF (International Monetary Fund), the United Nations, the European Economic Community, and others. This assistance will allow companies to get involved with very limited risk at these early stages of redevelopment.

If your company delays involvement in these early stages, you will only ensure the advantage for your competitors in establishing the market for themselves. Your delay at this stage in the development of the market will only serve as encouragement to improve competitions profitability while dramatically degrading your own position. If you take the initiative now, you will have that advantage and your competition will have to catch you in order to establish a position in the market and capture a market share. Time and understanding of these markets are critical while trying to establish contacts, build relationships, and develop a competitive position.

2.2 Countries in Focus

As discussed in the Executive Summary, the greatest current business opportunities are in Poland, Hungary, and Czechoslovakia. This report focuses on these three countries. Within the scope of this report, emphasis is not placed on East Germany, Romania, Bulgaria, or Yugoslavia.

2.3 Poland

The Republic of Poland is a democratic state occupying an area slightly smaller than New Mexico with a population of about 38 million. It shares borders with Czechoslovakia, Germany, Russia, and Belarus and has open access to the Baltic Sea. The terrain is mostly flat plains with mountains in the southern regions. Poland's natural resources are comprised of coal, sulfur, copper, natural gas, silver, lead and salt. It has ample arable land for agricultural crops with areas of timber and woodlands.

The economy is primarily agricultural with 75 percent of production from the private sector, the balance from state farms. Industrial development has been made on the Soviet model and held in the public sector. The programs to privatize the industrial sector have caused large but expected drops in production, unemployment and increased dissatisfaction with newly won freedoms and independence. These dislocations (expected as the biggest changes required to make the system work) are in the minds of the people.

The Solidarity led government took dramatic steps following their rise to power in 1990. They moved directly to a market-based economy. The Solidarity government took immediate steps to eliminate subsidies, free prices, establish a convertible currency (Zloty), and privatize the economy. These dramatic changes caused another new government to be elected in 1991. This new government has adopted and advanced most of the reforms of the previous government, with an effort to more rapidly privatize business and industry. The privatization is going to be a very difficult concept to install. Virtually none of the population working in the service or industrial sectors have any recollection of life under a free enterprise system. They have lived their entire working lives being taught not to demonstrate initiative or ambition. This is not completely true for the rural and agricultural sectors of these economies. These groups have been further removed from the centers of government power and control. They have also had to compete with the forces of mother nature and within the flourishing grey markets for goods and services. This group is a majority of the Polish population. It will therefore provide a basis for prospering under the new market-based economy.

The economic and trade policy reforms were adopted in January 1990. The program included a sharp reduction in state subsidies, elimination of nearly all price controls, legislation to eliminate monopolies, trade liberalization, currency convertibility, increased private ownership, and reform of the banking and tax systems. This market-oriented reform program was designed to gradually, but quickly, convert the former socialist system into a market economy. The success of the reform program is being assisted by IMF, the World Bank, international debt relief, and a $1 billion grant to help stabilize the Zloty. This program reduced inflation and shortages, but as would be expected it has increased unemployment and economic activity.

In July 1990, the Polish legislature passed privatization legislation to create publicly traded joint stock companies from state-owned enterprises. This legislation to encourage foreign ownership of Polish enterprises allows 100 percent ownership and repatriation of all after-tax profits.

These economic, trade and agricultural reforms eliminated price and production subsidies, price controls, and restrictions on procurement, retail operations and foreign trade.

Additional information and details can be found in Appendix A.

2.4 Hungary

Hungary is a landlocked country located in central Europe. It maintains a common border with Austria on the west, Czechoslovakia on the north, the Ukraine on the northeast, Romania on the east and several former Yugoslavian republics on the southern border.

Natural resources are not one of Hungary's strengths. It does have bauxite, coal, iron, uranium, oil and gas. Yet none of these resources is found in quantities sufficient for even domestic requirements. The greatest natural resource is the soil which supports a thriving agricultural industry.

The people, with their industrious nature, are one of Hungary's strong points. They have an independent nature which created production and trading in the agricultural sector even under the Soviet central plans. This productive nature developed trade and markets for Hungarian agriculture in parts of the world. In the industrial sector the desire to produce and trade, to create value, was not as free to grow as the agricultural sectors of the economy. The inadequate infrastructure, inept central management, and the emphasis on development of heavy industries stifled growth and the production of consumer goods and services. This misdirected emphasis on development of heavy industry has also inflicted severe environmental damage. Yet this strength is now the basis of the growth and re-emergence of an active, productive market-based economy.

Economic reforms have included foreign trade liberalization, price deregulation, and an expanded private sector role in the redevelopment of the country. Investment legislation allows 100 percent foreign ownership of Hungarian business, profit repatriation and liberal tax benefits and currency convertibility planned for 1993.

Agricultural and land reform was started in 1990, and farm prices were deregulated and consumer price subsidies were eliminated. Legislation in 1991 required that state and cooperatively owned assets be distributed to the original owners or they were to receive compensation according to established schedules. The privatization process has proceeded slowly and most production is still from state and cooperative farms.

The foreign trade monopoly was eliminated in 1990, allowing all qualified enterprises to directly engage in international trade. Lack of trade skills and experience, few contacts with foreign buyers and suppliers, and little knowledge of Western business practices have caused reliance on the former state-owned trading organizations.

Hungary, Czechoslovakia and Poland have agreed to create a trilateral Free Trade Area which may cover a large portion of trade among the three countries.

The U.S., under support of the East European Democracy Act of 1989, provides $60 million to establish the Hungarian - American Enterprise Fund and additional funds for other endeavors. This same legislation allows extension of Export-Import Bank credits and the insurance of U.S. investment under OPIC. Waiver of the Jackson-Vanik Amendment has allowed USDA export credit guarantees since 1978. In 1989, continuous MFN (Most Favored Nation) status was granted to Hungary, with provision for semiannual certification [6].

Additional information can be found in Appendix B.

2.5 Czechoslovakia

The Federation of Czechoslovakia was composed of the republics of Czech and Slovak. The Czech and Slovak Federal Republics (CSFR) implemented market-oriented economic reform in June 1990. These reforms were implemented under the political alliance forming a coalition government with Vaclav Havel as president and Marian Calfa as prime minister. Current political action has resulted in the separation of these two republics. The Czech Republic is the most industrialized and most populous. The Slovak Republic, the eastern portion of the former federation, is more rural and less populated.

In February 1990, President Bush announced a waiver of the Jackson-Vanik Amendment for the Czech and Slovak Republics. This action makes the countries eligible for a variety of U.S. government credit and credit guarantee programs. In 1991, the republics were granted continuous MFN (Most Favored Nation) status, subject to a semiannual certification requirement. In December 1991, the president signed legislation that gives him discretion to determine that Title IV of the Trade Act of 1974, including the Jackson-Vanik Amendment, should no longer apply to the republics.

In November 1990 a three-year bilateral trade relations agreement was implemented, which confers reciprocal MFN tariff treatment. In 1991, both countries signed a bilateral investment treaty which provides basic guarantees to U.S. investment in the republics and established the Czechoslovak-American Enterprise Fund, with initial funding of $5 million, to assist the CSFR private sector.

The Trade Enhancement Initiative for Central and Eastern Europe was launched by the U.S. government in March 1991 to promote exports by the region.

The CSFR and the European Communities (EC) have signed a economic trade agreement in 1990 to expand bilateral trade. In November 1991 they concluded an Association Agreement aimed at gradually phasing in a bilateral Free Trade Agreement.

General Agreement on Trades and Tariffs (GATT) has granted the republics a blanket waiver to raise previously bound tariffs on agricultural products.[7]

The U.S. and Foreign Commercial Service, Prague and the Eastern Europe Business Information Center suggest the best prospects for trade and investment during 1991 were:

- Computers and Peripherals
- Pollution Control Equipment
- Computer Software and Services
- Telecommunications Equipment
- Telecommunications Services
- Electrical Power Systems
- Information Services
- Education and Manpower Training Services
- Management Consulting Services
- Financial Services
- Travel and Tourism Services
- Aircraft and Parts
- Avionics
- Medical Equipment
- Franchising
- Accounting Services
- Drugs and Pharmaceutical
- Food Processing and Packaging Equipment
- Hotel and Restaurant Equipment
- Renewable Energy Equipment

The Czechoslovakian economy has been one of the most developed in Central and Eastern Europe. Yet the adherence to the ideal of central planning and the development of heavy industry along the Soviet model impeded real productivity. This uncontrolled development has degraded the environment in some areas of the country to almost lethal levels.

"The Czechs have yet to finish their environmental analysis. What are the worst problems? Everything. The air, the water and the land." Jiri Nechvatal of the State Environmental Committee shrugs his shoulders.

Additional information can be found in Appendix C.

CHAPTER 3

BUSINESS OPPORTUNITIES

There are business opportunities beyond comprehension in these countries. These are markets that require nearly everything to put these 1960 economies back on their feet in the 1990's. The equipment, systems, engineering, design, and technology are not readily available or understood by the officials and technicians in these Eastern European economies. They are ready to make that 30 year leap to the efficiency and technology that is available to further reduce costs and pollution of the environment.

It would appear that Western European firms would have a natural advantage in gaining access to these markets for development and sales. Yet this doesn't seem to be the case. Each of these countries has made a concerted effort to maintain a broad base of outside involvement, investment income and technology. There is also a very strong desire to encourage heavy U.S. involvement in their new development as they move directly from the 1960's to the 1990's.

The key for U.S. companies will be to identify a position in this development process where their products, technology and services will provide a defined benefit. Does that sound familiar? The biggest surprise to U.S. businessmen and their companies will be that doing business there is no different than doing business here at home. You have to provide the product and service to fill an identified need, then show the potential buyers the benefit. The added problems of language, custom, currency and distance are not nearly as great as may be thought. If the businessman concentrates on the solution of the targeted dilemma, those obstacles such as language and financing will only be details to be dealt with in the solution of the larger problem. It is surprising how easy this business can be if we can get beyond the intermediate stumbling blocks and deal with the real problems to be solved.

Although the areas of opportunity are almost limitless, some of the identified sectors of the economy are energy, environment, transportation, agriculture, manufacturing, telecommunications, health and medical care, financial services, and computer and information services, to name only a few possible areas of concentration.

3.1 Energy

3.1.1 Generation

Electrical power generation has been one of the major contributors to the severe air, water and soil pollution throughout the region. Lignite coal is plentiful throughout the region and became the fuel of choice as it was inexpensive. More recently, electrical generation has been built using nuclear power, import oil and gas. The imported oil and gas subsidies have now disappeared and there is a movement to upgrade or close the Soviet-design nuclear reactors.

Many of the lignite coal-powered plants are over 25 years old, and the nuclear plants are of Soviet design and require upgraded control and safety systems. Reduced industrial production has reduced demand for electrical power. A reduction of 20 to 30 percent will allow sufficient capacity to meet all needs for the next several years.[8] Most of the existing equipment and facilities will have to be replaced over the next 10 years. The equipment and systems that are not replaced will have to employ new control and management systems to improve efficiency in operation and fuel utilization, while reducing polluting emission in the environment.

Lignite coal has been a major contributor to the severe environment problems that have overtaken entire regions of each of these countries. The demand for cleaner fuels and reduced environmental pollution is creating a huge demand for technology, equipment and expertise to reduce pollution. This demand, along with the technology and equipment to upgrade systems and safety for the region's nuclear power plants, will generate great opportunities.

3.1.2 Transmission/Distribution

Transmission and distribution in Czechoslovakia has accomplished completely separate grids in Czech and Slovak republics which total almost 14,000 kilometers. Most of the system and equipment is old and inefficient. Operation and maintenance costs for the system are high and reliability is declining. Almost a complete replacement of the transmission systems will be required through the decade ahead to support redevelopment of industrial production and increased domestic demand following the turn of the century.

In Hungary work is already underway to determine requirements to upgrade and standardize the power transmission grid and then expand the grid for connection to the Western European Grid by 1995.[9]

The Polish Power Grid Company maintains over 12,000 kilometers of transmission lines, most of which are over 30 years old. There are plans to expand the grid in the northeast, with the addition of substations and connection to the Western European Grid. This is thought to be 5 to 6 years in the future.

Additional information and details concerning these systems can be found in the "NEMA Fact-Finding Delegation Report on the Electric Power Transmission and Distribution Sectors in Czechoslovakia, Hungary and Poland" (Appendix U) and "Trip Report - Eastern Europe," W.H. Mashburn, Mechanical Engineering Department, Virginia Tech, July 1991 (Appendix V).

3.1.3 Equipment/Efficiency

Industry and manufacturing will be required for reasons of competition. For environmental improvement, more efficient energy use will also be required in the very near future. High efficiency boilers, energy handling and distribution systems, motors and equipment will be one of the biggest opportunities. This market will be huge for those who become actively involved. Those who wait for the buyers to show up at their door will be very disappointed when they see their competition prosper with accounts that could have been theirs.

The opportunities will not only be in the industrial sectors. The business and residential users of energy for heat, light and equipment will also be huge markets with major dividends to be paid to the companies involved and to the countries, people and economies as beneficiaries. An example is the heating of business, office and residential structures in Warsaw and other Eastern European cities. The heavily subsidized energy costs of the past made concern for improved use and efficiency almost non-existent. This lack of control is demonstrated by the almost total lack of any thermostatic control of heating systems. The heat was turned on according to the time of year and then controlled by opening windows to allow excess heat to escape. This was not an effort to modify the local climate, but simply an example of lack of understanding and concern. This example of basic absence of control at the most elementary level is the standard rather than the exception.

3.2 Environmental Conditions

3.2.1 Conditions in Brief

Czechoslovakia:
- 70 percent of rivers heavily polluted,
- 40 percent of sewage untreated,
- Half the forests are dying or damaged,
- Pollution-related cancers and infant mortality rates are soaring.

Hungary:
- Drinking water in the south seriously contaminated with arsenic,
- Sewage treatment in poor state everywhere,
- Air pollution exceeds maximum limits for half the population.

Poland:
- 95 percent of rivers polluted,
- High acid levels and metals pollution have damaged half the farmland,
- In Upper Silesia, half of children suffer pollution-related illnesses.

Bulgaria:
- One third of forests are damaged,
- Black Sea polluted by sewage, oil, industrial waste,
- Metals pollution destroying farmland soils.

East Germany:
- Urban air pollution 50 times higher than national standards,
- In some areas 90 percent of children suffer respiratory diseases,
- Open uranium waste dumps polluting groundwater and air in the south.

Romania:
- Virtually no air pollution controls,
- High levels of pollution-related heart disease and infant mortality,
- Toxic waste disaster at the port of Sulina on the Black Sea.[10]

The examples above are only a sampling of the myriad of documented problems that face the region.

Environmental conditions throughout the region are a disaster affecting the air, water, land, crops, wildlife and the people. The problems in some of these areas have been called the worst known pollution problems in the world. The Elbe River flows from Czechoslovakia through Germany and is so polluted that those who fish there take a major health risk. This river was the border between East and West Germany. It was said that those trying to escape who were not shot would later die of poisoning from exposure to the river water.

"The Czechs have yet to finish their environmental analysis. What are the worst problems? Everything. The air, the water and the land. Jiri Nechvatal of the State Environmental Committee says at least his country was largely too poor to buy pesticides. In Poland, officials concede they do not have the $400 million it wold take to halt environmental damage, much less the $25 billion they believe is needed to reverse it."[11]

For 40 years the Czechoslovakian government kept an iron hand on any information concerning the environment. No negative reporting was allowed. The only concern was to keep the industrial engine running, whatever the cost. The cost is that life expectancy is now 10 years less than in European cities to the west. Last fall in Most, a Czech industrial town not far southwest of Cecin, the air was so poisoned that children had to wear breathing masks during their walk to school. Allergies and respiratory diseases are rampant there. Cancer killed nearly a third of the population, including many children. A local environmentalist said that every family had lost at least one member to the pollution. There are countless other examples. Leaking sewage has contaminated a third of the drinking water. Beyond that, toxic wastes have killed a third of the region's inland waters. Czechoslovakia has chosen to keep its dangerously antiquated nuclear reactors in operation rather than increase its use of lignite.[12]

A lake near a chemical plant in the town of Kemerovo, 3,000 km east of Moscow, is so polluted that citizens of the town dispose of stray dogs by throwing them into the water. In addition, the water is so contaminated with phenols that the carcasses dissolve within days. In some parts of Eastern Europe highly polluted air and water are blamed for deformed babies, elevated child mortality rates and a wide variety of illnesses among adults. Indeed, after decades of Communist rule, large areas of the Soviet Union and Eastern Europe have been reduced to ecological wastelands. Now efforts are underway to correct the damage.[13]

There's worry in Western Europe, too, where pollution from the east is killing forests and poisoning the Baltic Sea. Of the $365 million the European Commission pledged to Poland and Hungary in 1990, up to 20 percent is earmarked for the environment. Denmark's parliament just voted $75 million in ecological aid for Eastern Europe, and Sweden is giving an additional $180 million. But it's not enough. Cleaning up Poland's pollution will cost $20 billion, and other Eastern European states need billions, too. "We know what we need to do," says Boldrich Moldan, the new environment minister for the Czech Republic, the larger of Czechoslovakia's two states, "but we don't have the money to do it."

This shortfall will make it doubly hard to repair the region's economy. The World Bank estimates that 10 percent of Poland's gross national product is lost because of illness and industry's inability to use polluted water. "For years, experts have said that environmental degradation ultimately affects economic growth," says Thierry Baudon, the World Bank's chief environment official for Europe. "In Eastern Europe, you can take the 'ultimately' out of that statement."[14]

Yet one of the big environmental problems facing each of the Eastern European countries will be the cleanup of the former Soviet military installations, which dot the entire region with hundreds of sites. There are stories of dumpsites and disposals that could raise genuine fear in anyone hearing them. In Czechoslovakia, it is reported there are 8,000 square miles that have been polluted with dumped fuel, outdated munitions, and discarded biological and chemical weapons in unmarked dumps and unmapped sites. At a site in northern Morava, departing Soviets emptied 15,000 gallon fuel tanks on the ground, in order to transport the tanks home. Now the groundwater beneath the depot is covered with 8 to 10 inches of diesel fuel, and the aquifer is so saturated with oil that when it rains local streams have an oil sheen on them. "You could practically drill for diesel there," says Vlcek. Mayor Josef Bifus of Vysoke Myto, a picturesque town in central Bohemia, says that drill testing of groundwater since the Soviets began to leave has indicated contamination "30 to 50" times allowable levels for nearly everything.[15] They estimate that it will cost $125 million in the next few years, but other sources feel this in only a small fraction of what the real cost would be. In another case an environmental consultant was inspecting one of the former Soviet military installations. He mistakenly stepped in a small pool of black material, which dissolved his shoe in the next few minutes. In the city of Prague, Adam Turner says of the air pollution, "This is the worst I've ever seen it anywhere in the world." Forty years of uncontrolled automobile exhaust and heavy industrial pollution have created grave dangers for Czechoslovakia, and now, it's dramatically limited visibility across the nation's capital.[16]

These are but a few examples of the problems that are only now being discovered throughout the region. Some who feel that the worst is yet to be realized as the cleanup begins and the long-term health problems become more evident.

3.3 Waste Management

Waste management is an almost unheard of science and industry in these countries of Eastern Europe. Uncontrolled dumping and disposal of everything from solid waste to hazardous material and toxic chemicals have created a witches' brew that overshadows the entire region. Everything is somehow affected: from the air in the cities, towns and industrial areas; to the groundwater and aquifers that have been polluted with heavy metals, petroleum products and toxic chemicals; to the poisoning of the soil in forests and farmlands due to dumping and the settling of air born particles.

This is a whole new arena in which development and control will have to start with the use of western technology and expertise to determine the type and degree of the environmental pollution. This is an effort that will require the expenditure of tens of millions of dollars before anyone is in the position to effectively define and tackle this problem.

3.4 Pollution Control

Pollution control programs, technology, equipment and expertise of all descriptions are needed in these countries. Virtually nothing is available and therefore everything is needed.

Scrubbers are required in industrial plants and power plants to remove emissions from the burning of lignite. Improving the efficiency of power utilization will reduce the total requirement for power and therefore reduce total emissions. Automobile pollution control systems and improved engines will be required. Fuel reprocessing is to provide cleaner fuels and new fuels so the burners and the systems can use these new fuels. The needs are endless, as are the opportunities for those individuals and companies with the products, technology, insight and the determination to get involved and stay involved.

3.5 Manufacturing

Most manufacturing facilities are from the massive industrialization of the Soviet Union in the 1950's and 1960's. These facilities were built with only products in mind and no concern for the control of pollution from emissions, effluent or disposal of hazardous or toxic wastes. These facilities will have to be upgraded with new systems, equipment and controls to improve efficiency and waste control. These manufacturing facilities have immediate need for almost anything that can be imagined. Again, the opportunity is limited only by the imagination and the ability of the U.S. firms and (1) the companies in the targeted countries and industries in the purchase and incorporate the newer technology on control (2) the willingness of companies and industries in the targeted countries to purchase and incorporate newer pollution control technology.

CHAPTER 4

FEDERAL PROGRAMS

4.1 U.S. Government

One of the charges of the U.S. government is to assist U.S. companies and citizens in the promotion and establishment of U.S. products, goods and services in markets around the globe. Several departments and agencies have developed specialized staff and programs to assist U.S. companies and citizens to identify and develop these new or expanded overseas markets. These programs and services are designed to provide information and assistance to the businessman who is only starting to think about participation in the new global marketplace, as well as to the veteran international marketer who may see new or expanding market opportunities.

The Departments of Agriculture, Commerce, Treasury, State and the International Trade Commission, the Small Business Administration, and the Office of the U.S. Trade Representative are only a few of the U.S. government departments and agencies that work diligently to help you identify, develop and expand markets for your U.S.-produced products, goods and services.

4.2 U.S. Department of Commerce

The U.S. Department of Commerce maintains agencies such as the U.S. & Foreign Commercial Service, the International Trade Administration and specialists stationed in Washington, U.S. Regional Offices and in each of the targeted countries. These knowledgeable individuals are available to assist you and your company in understanding the market, establishing contacts, and negotiating the unfamiliar maze of foreign legislation, customs and the national system in general.

In Warsaw, Poland, they maintain an "American Business Center" where you will be able to base you efforts while in the country. The can provide office facilities and information services, telephone, FAX and computers for very reasonable rates. Rates and additional information can be found in

Appendix O.

4.3 U.S. Department of Agriculture

The U.S. Department of Agriculture (USDA) has agencies such as the Foreign Agricultural Service (FAS), the Food Safety Inspection Service (FSIS), the Federal Grain Inspection Service, (FGIS), Economic Research Service (ERS), American Marketing Service (AMS), Agricultural Attach Service, and many others that will be defined in greater detail later. The specialists and the programs provided by this agency are for the benefit of U.S. agricultural producers, processors, distributors and all others that are involved with the production, processing, and marketing of U.S. food, feed and fiber.

Additional information is presented in Appendix W.

CHAPTER 5

BANKING AND FINANCE

5.1 Export-Import Bank

The Export-Import Bank of the United States (Eximbank) is an independent U.S. government agency that helps finance the sale of U.S. goods and services to foreign buyers. Over the past 58 years, Eximbank has helped finance more than $220 billion in U.S. exports.

Eximbank supports export sales by making it easier for overseas buyers to purchase U.S. goods and services on credit. It does this in two ways: 1) Eximbank provides loans directly to foreign buyers; 2) It encourages U.S. suppliers or their banks to extend credit terms to foreign buyers by providing credit risk protection and, in some cases, fixed-rate funding support. Eximbank also helps U.S. exporters obtain pre-export financing through its Working Capital Guarantee Program.

Information on the Export-Import Bank is presented in Appendix Y.

5.2 Hard Currency Availability

The availability of hard currency in each of the targeted countries has become a reality since they have each declared their independence. Each of the targeted countries has developed legislation for the free movement of currency, the convertibility of their own currency with International Monetary Fund (IMF) assistance, the freedom to repatriate after tax profits and earning, and other programs and legislation critical to the development of investment and economic growth.

5.3 Currency Conversion

The IMF has worked with the governments of each of these new countries to support the convertibility of their currency. In most cases this is done through bimonthly auctions. These auctions will establish a rate and eventually trading of the currencies at values based on the auctions and then later general trading.

5.4 Taxation

The taxation of foreign companies, investments, and the real property and plants owned by foreign companies and joint-ventures is being liberalized. New legislation is being developed and introduced in each of these countries on an ongoing basis in these earlier stages of change. The best advice would be to have the various U.S. government agencies that monitor these changes provide up-to-date information as it is required for your business and investment needs.

5.5 U.S. Tax Treaties

Bilateral trade agreements are now being negotiated or updated with each of these countries. It can be expected that tax treaties will be negotiated as one of the next steps. Again, the best advice would be to have the various U.S. government agencies that monitor these changes provide up-to-date information as it is required for your business and investment needs.

5.6 Trade Financing Programs

Sources for financing ventures in Eastern Europe are presented in Appendix X.

CHAPTER 6
INFORMATION RESOURCES

6.1 Country Reports

6.1.1 USAID (U.S. Agency for International Development)

6.1.1.1 Domestic Environmental Products and Services Sectors: Czechoslovakia, An Accessment of the Current Status and Future Potential.

6.1.1.2 Domestic Environmental Products and Services Sectors: Hungary, An Accessment of the Current Status and Future Potential.

6.1.2 U.S. Department of Commerce

6.1.2.1 Marketing Plan-Poland (see Appendix J)

6.1.2.2 Marketing Plan-Hungary (see Appendix K)

6.1.2.3 Marketing Plan - Czechoslovakia (see Appendix L)

6.1.2.4 Key Contacts: Poland (see Appendix D)

6.1.2.5 Key Contacts: Hungary (see Appendix E)

6.1.2.6 Key Contacts: Czechoslovakia (see Appendix F)

6.2 Environmental Bibliography

6.2.1 Eastern Europe

6.2.1.1 "East European Pollution - DIRTY STORIES", The Economist, February 1, 1992, page 51.

6.2.1.2 "The West Must Help a Poisoned East", Business Week, March 19, 1990, page 146

6.2.1.3 "East Europe's Dark Dawn", National Geographic, June 1991, page 42-63.

6.2.1.4 "Where Night Falls All Day Long", National Geographic, June 1991, page 64-68.

6.2.1.5 "Dark Days, Science and the Citizen", Scientific American, August 1990, pages 16&20.

6.2.1.6 "East Europe Floundering in Sea of Pollution", East European Diary, Tim McCarthy, National Catholic Reporter, February 1,1991, page 16.

6.2.1.7 "The Cost of Neglect", Environment, McCleans, May 7, 1990, pages 54-55.

6.2.1.8 "Eastern Europe's Big Cleanup, Science and Technology, Environment", Business Week, March 19, 1990, pages 114-115.

6.2.1.9 "Eastern Europe Faces Crisis, Pollution", USA Today, Newsview, April 1989.

6.2.2 Czechoslovakia

6.2.2.1 "Cleaning up After the Soviet Army", Peter S. Green in Czechoslovakia, U.S. New and World Report, May 28, 1990, page 40.

6.2.2.2 "Tennessean Cleans up in Prague", Adam Turner, Tennessean, February 7, 1991.

6.2.2.3 "Where the Sky Stays Dark", Frederick Painton, Environment, Time, May 28, 1990, page 40-41.

6.2.2.4 "Czechoslovakian Landscapes", Sam Bingham, Audubon, January 1991, page 92(12).

6.2.2.5 "A New Dawn in Bohemia?", Rob Waters, Sierra, May-June 1990 page 34(6).

6.2.2.6 "So Long, Ivan - Don't Bother to Write", Gail Shares, Business Week, October 8, 1990, P28A(2)

6.3 Other Reports

6.3.1 U.S. Department of Commerce

6.3.1.1 East European Business Bulletin
This is a bimonthly newsletter and information publication produced by the Eastern Europe Business Information Center, U.S. Department of Commerce, International Trade Administration. (see Appendix Q)

6.3.1.2 East Europe Looks for Partners Information, trade leads and commercial development information concerning the countries of Eastern Europe. (see Appendix N)

6.3.1.3 Business America - The magazine of International Trade, Publication of the U.S. Department of Commerce, May 18, 1992. (see Appendix M)

6.3.1.4 "The National Trade Data Bank", The Export Connection, CD-ROM compact disc information. (see Appendix R)

6.3.1.5 U.S. Department of Commerce Business Services In Eastern Europe. Eastern Europe Business Information Center (EEBIC).

6.3.1.6 Central & Eastern Europe Business Information, Listing of Reports Covering: Trade, Markets, Finance, Investment, Joint Venture Leads. Reports by International Trade Administration (ITA), Eastern Europe Business Information Center (EEBIC). (see Appendix P)

6.3.1.7 EEBICFLASH

This is a 24-hour automated FAX delivery system which will provide a variety of documents and information for each of the countries of Eastern Europe and the Baltics. (see Appendix S)

6.3.2 Association Reports

6.3.2.1 "Business Opportunities in Eastern Europe For Energy Efficient Industrial Products", The Alliance to Save Energy, 1992. (see Appendix T)

6.3.2.2 "Fact Finding Delegation Report on the Electric Power Transmission and Distribution Sectors in Czechoslovakia, Hungary and Poland", The National Electrical Manufacturers Association, July 1991. (see Appendix U)

6.3.2.3 Trip Report - Eastern Europe, W.H. Mashburn, Mechanical Engineering Dept., Virginia Tech July 1991, for The Association of Energy Engineers. (see Appendix V)

6.3.3 World Bank

Environmental and Energy Division
Mr. Bernard Montfort
Division Director
World Bank
1889 H Street, N.W.
Washington, D.C.
202/473-2700

Poland
Mr. Busz
Room # H12003
World Bank
1889 H Street, N.W.
Washington, D.C.
202/473-2686

Czechoslovakia
Mr. Grey
Room # H12017
World Bank
1889 H Street, N.W.
Washington, D.C.
202/473-2692

Hungary
Mr. Markus
Room # H12007
World Bank
1889 H Street, N.W.
Washington, D.C.
202/473-5209

6.3.4 Overseas Private Investment Corporation (OPIC)

OPIC is a self-sustaining U.S. government agency whose purpose is to promote economic growth in developing countries by encouraging U.S. private investment in those nations. (see Appendix Z)

OPIC assists U.S. investors in this effort through two principal programs: (1) The insurance of investments against certain political risks and (2) the financing of such investments through direct loans and/or loan guarantees. All of OPIC's insurance and guarantee obligations are backed by the full faith and credit of the United States of America, as well as OPIC's own substantial financial reserves.

For American businesses considering overseas ventures, obtaining basic information about foreign countries and their business environments is an important first step. To assist U.S. firms in gathering such information, as well as facilitate the flow of information about developing countries to potential U.S. investors, OPIC has created the Investor Information Service (IIS). IIS is a publications clearinghouse that provides interested companies and individuals with easy "one-stop-shopping" for basic data and information commonly sought when considering investment overseas.

The materials, which are gathered into "kit" form, have been obtained from various U.S. government agencies, foreign governments and international organizations. Together their source materials cover the economies, trade laws, business regulations, attitudes, political conditions, and investment incentives of specific developing countries and areas.

For further information about OPIC, its programs and services, write or phone:

Information Officer
Overseas Private Investment Corporation
1615 M Street, N.W.
Washington, D.C. 20527
202/457-7010
800/424-6742 (outside the D.C. area)

6.3.5 Private Export Funding Corporation (PEFCO)

PEFCO is a private corporation owned primarily by a group of commercial banks. PEFCO works with Eximbank in using private capital to finance U.S. exports. Their loans must be covered by Eximbank's unconditional guarantee. PEFCO generally makes loans on a participation basis, with their participation being no less than $1 million. The buyer should approach PEFCO in conjunction with a traditional lender or with the U.S. supplier.

For further information, contact:

Private Export Funding Corporation
280 Park Avenue
New York, NY 10017
212/557-3100

6.3.6 EXIMBANK

The Export-Import Bank of the United States is the U.S. government agency that helps finance export sales of American goods and services. In its 51 years of operation, Eximbank has used loan, guarantee and insurance programs to support more than $170 billion in U.S. exports. (see Appendix Y)

Eximbank has an agency and reinsurance agreement with the Foreign Credit Insurance Association (FCIA), a group of U.S. property, casualty and marine insurance companies that sell and service its export credit insurance policies.

The interest rates on Eximbank's Direct Loan Program, the Engineering Multiplier Program, the Medium-Term Credit Program and the Small Business Credit Program are determined by the classification of the country to which the export will be shipped.

Contact Eximbank for full program details and application instructions. Call Eximbank's Advisory Service at 800/424-5201: Washington, D.C., Alaska and Hawaii should call 202/566-8860, or write:

Marketing Department
Eximbank
811 Vermont Avenue, N.W.
Washington, D.C. 20571

6.4 Contact Base

6.4.1 Government Agencies

US DEPARTMENT OF AGRICULTURE

Commodity Credit Corporation
US Dept. of Agriculture
14th & Independence, SW
Washington, DC 20250
(202) 720-4785

Federal Grain Inspection Service
US Dept. of Agriculture
14th & Independence, SW
Washington, DC 20250
(202) 457-3367

Food Safety Inspection Service
US Dept. of Agriculture
14th & Independence, SW
Washington, DC 20250
(202) 720-8217

Agricultural Attache Service
US Dept. of Agriculture
14th & Independence, SW
Washington, DC 20250
(202) 720-3253

Foreign Agricultural Service
US Dept. of Agriculture
14th & Independence, SW
Washington, DC 20250
(202) 720-3448

Commodity Marketing Divisions
Grain and Feed Division
Dairy, Livestock and Poultry Division
Forest Products Division
High Value Products Division

Oilseeds and Products Division
Horticultural and Tropical Products Division
Tobacco, Cotton and Seeds Division

(Contact name and number for each of the divisions listed above are available through the listing for the Foreign Agricultural Service.)

U.S. GOVERNMENT PRINTING OFFICE

Superintendent of Documents
Government Printing Office
Washington, DC 20402
(202)783-3238

BASIC GUIDE TO EXPORTING

The Basic Guide to Exporting provides you with a detailed, step-by-step guide through the exporting process. It fills you in on everything you need to know about developing an export strategy, finding economical market research, pricing for profits, shipping overseas, export documentation, traveling abroad, answering overseas inquiries, making an overseas sales contact, using an export management company, antitrust, copyrights, trade secrets, technology licensing, getting paid for overseas sales, joint ventures, patents, licensing exports, financing overseas sales, foreign sales corporations, government assistance to exporters, and a wealth of other useful information. It is available for $8.50
(s/n 003-009-00487-0) from GPO.

COMMERCE BUSINESS DAILY

Commerce Business Daily (CBD) is a daily list of government procurement invitations, contract awards, subcontracting leads, sales of surplus property and foreign business opportunities, as well as certain foreign government procurement. The annual subscription cost is $261.00 for first class mail and $208.00 second-class. A six-month trial subscription costs $130.00 for first-class and $104.00 second-class. CBD is available on-line from a sizable list of contractors named in the lower left-hand corner of the front page of each edition. Contact ITA for additional information, or GPO to order a subscription.

COUNTRY & TERRITORY INFORMATION PAMPHLETS

Background notes, a series of short, factual pamphlets about various countries and territories of the world, plus selected international organizations, contain up-to-date information on each country's people, culture, geography, history, government, political conditions, economy, defense, and foreign relations with other countries, including the United States. A reading list provides additional sources of information about the country, and travel notes, maps, and occasional photographs are included. A complete set can be purchased from the Government Printing Office for $32.00.

EXPORT TRADE RESOURCE DIRECTORY

Partners in Export Trade is a directory for export trade contacts published by the U.S. Department of Commerce, containing a list of over 4,500 firms, including many ETCs and EMCs, manufacturers, producers, and service organizations. It is available for $11.00 from GPO (s/n 003-009-00512-4).

EXPORT TRADING COMPANY (ETC) GUIDEBOOK

The Export Trading Company (ETC) Guidebook is intended to assist those who are considering starting or expanding through the various forms of an ETC as encouraged by the ETC Act. This guidebook will also facilitate your review of the ETC Act and export trading options and serve as a planning tool for your business by showing you what it takes to export profitably and how to start doing it. It is available for $8.00 (s/n 003-009-00523-0) from GPO.

FOREIGN ECONOMIC TRENDS

Foreign Economic Trends presents business and economic developments and the latest economic indicators for almost 100 countries. They are prepared on an annual or semiannual basis by the U.S. Foreign Service. More than 100 reports are issued each year. Annual subscription is $55.00 from the Superintendent of Documents or single copies are available for $1.00 from the Department of Commerce, Publications Distribution, H161M, Washington, DC 20230 or (202) 377-5494.

HOSTAGE: PREPARATION/AVOIDANCE/SURVIVAL

The State Department's Bureau of Diplomatic Security publishes this pamphlet which explains what you can do to make yourself less susceptible to terrorist violence and how to improve your chances of survival should your efforts fail. Information includes personal preparations, security measures, and tips on hostage survival.

U.S. EXPORT ADMINISTRATION REGULATIONS

The Export Administration Regulations is a comprehensive guide to the rules controlling exports of U.S. products, and it answers questions on export licensing requirements. At no additional cost, subscribers receive supplementary Export Administration Bulletins which explain recent policy changes and provide replacement pages to keep your set of regulations up-to-date.

OVERSEAS BUSINESS REPORTS

Overseas Business Reports include current and detailed marketing information, trade outlooks, statistics, regulations, and marketing profiles for most countries. The annual subscription from GPO is $14.00, and $17.50 for foreign mail orders. Single copies are available from the Department of Commerce, Publications Distribution, H1617M, Washington, DC 20230; (202) 377-5494

US DEPARTMENT OF COMMERCE

Department of Commerce
14th & Constitution, NW
Washington, DC 20230

EASTERN EUROPE BUSINESS INFORMATION CENTER
U.S. Department of Commerce
14th & Constitution, NW, Room 6043
Washington, DC 20230
(202) 377-2645

This office serves as a clearinghouse for information on opportunities in Eastern Europe, and on U.S. government programs supporting private enterprise, trade and investment in Eastern Europe. It is also a referral point for voluntary assistance programs. It will encourage development of economically sound proposals for the Polish-American and Hungarian-American Enterprise Funds.

EXPORT ADMINISTRATION PUBLICATIONS
Export Seminar Staff
Office of Export Licensing
Bureau of Export Administration
Department of Commerce
14th & Constitution, NW, Room 1608
Washington, DC 20230
(202) 377-8731

The following free publications are available Through the Export Seminar Staff:

- -How to Conduct a Systems Analysis
- -How to Prepare for an OEL Systems Review
- -Pre-License Consultation
- -The Export License: How to Fill Out the Application-A step-by-step guide using Form BXA-622P
- -U.S. Antiboycott Laws
- -Regulation of Participation in Foreign Boycotts
- -Inter-Agency and Inter-Government Review Procedure

The following publications are for sale from:

Superintendent of Documents
Government Printing Office
Washington, DC 20402
(202)783-3238

-Export Administration Annual Report
-Annual Foreign Policy Report to Congress
-U.S. Export Administration Regulations

The Export Administration Regulations is a comprehensive guide to the rules controlling exports of U.S. products, and it answers questions on export licensing requirements. At no additional cost, subscribers receive supplementary Export Administration Bulletins which explain recent policy changes and provide replacement pages to keep your set of regulations up-to-date.

EXPORT ADVICE HOTLINE - (800) 342-4300

For information and advice on exporting your products or services, call this hotline and ask for Operator 940. This operator will take your name and address, and a U.S. Department of Commerce trade specialist will contact you.

EXPORT TRADING COMPANY ACT of 1982

Office of Export Trading Company Affairs
International Trade Administration
U.S. Department of Commerce
14th & Constitution, NW
Washington, DC 20230
(202) 377-5131

The Export Trading Company Act of 1982 is an information packet containing fact sheets on some of the ways Title III of the ETC Act can help U.S. firms gain a competitive edge in foreign markets. Topics include shipper's associations, trade associations and the ETC Act, technology licensing and the ETC ACG, and the Webb Pomerane Act and the ETC Act. For a copy, contact this office.

EXPORT TRADING COMPANIES

Office of Export Trading Company Affairs
International Trade Administration
U.S. Department of Commerce
14th & Constitution, NW
Washington, DC 20230
(202) 377-5131

If you represent a firm looking to expand sales by exporting, you can hire an Export Trading Company (ETC) or Export Management Company (EMC) to market and sell your products for you and save you time and money. The ITA sponsors these companies through conferences, workshops, and presentations. Some of the expertise and other advantages offered by ETCs and EMCs include the following:

Services Offered: Many ETCs and EMCs provide a multitude of services, including market research; appointing overseas distributors or commission representatives; exhibiting a client's products at international trade shows; and handling financing, shipping, and documentation. All the U.S. manufacturer has to do is fill the order. Most ETCs and EMCs are small firms which provide personal one-on-one service.

Market Knowledge: Their ability to identify and understand the diversities inherent with different markets is almost infinite. Regional traders cover many markets and geographical areas and help ETCs and EMCs adjust their services according to particular environments.

Product Stabilization: Most successful ETCs and EMCs specialize in one or more product areas. Both their product expertise and, more importantly, their knowledge of distribution structures in foreign markets focus on the specific interest of their clients.

Efficiencies of Scale: ETCs and EMCs have specialized personnel with international backgrounds, who use resources that are generally available only to the export departments of larger firms. However, small firms are not the only ones that benefit from efficiencies of scale--anyone can profit whenever an ETC or EMC can efficiently export a product at a lower cost than the manufacturer. Some benefits of using an ETC/EMC include language expertise, experience in financial and bank relations, and familiarity with regulations of importing countries.

Shipping Efficiencies: ETCs and EMCs can consolidate the products of several suppliers in one shipment, frequently negotiating better freight rates than most individual manufacturers can obtain for themselves.

There are more than 1,500 ETCs and EMCs in the United States. Potential exporters should develop a list of those ETCs and EMCs which specialize in exporting the types of products proposed to be sold overseas. For assistance in locating and selecting the proper ETC/EMC, contact your nearest ITA district office, or contact the office mentioned previously.

INTERNATIONAL TRADE ADMINISTRATION
U.S. and FOREIGN COMMERCIAL SERVICE
DISTRICT OFFICES

ALABAMA

*Birmingham	Berry Building, Room 302	
	2015 2nd Ave., North	(35203)
	Tel: 205/731-1331	

ALASKA

Anchorage	P.O. Box 32	
	222 West 7th Ave.	(99513-7591)
	Tel: 907/271-5041	

ARIZONA

Phoenix	Federal Building & U.S. Courthouse	
	Room 3412	
	230 North 1st Ave.	(85025)

ARKANSAS

Little Rock	Savers Federal Bldg., Suite 811	
	320 West Capital Ave.	(72201)
	Tel: 501/378-5794	

CALIFORNIA

Los Angeles	11777 San Vicente Blvd.,	
	Room 800	(90049)
	Tel: 213/209-6703	
Santa Ana	116-A W. 4th Street,	
	Suite #1	(92701)
	Tel: 714/836-2461	
San Diego	6363 Greenwich Dr.,	
	Suite 260	(92122)
	Tel: 619/557-5395	
San Francisco	Federal Bldg., Box 36013	
	450 Golden Gate Ave.	(94102)
	Tel: 415/556-5860	

COLORADO

*Denver	1625 Broadway, Suite 600	(80202)
	Tel: 303/844-3246	

CONNECTICUT	
*Hartford	Room 610-B, Federal Office Bldg. 450 Main Street (06013) Tel: 203/240-3530
DELAWARE	Serviced by Philadelphia District Office
DISTRICT OF COLUMBIA	
+Washington	Room 1066 HCHB, Department of Commerce 14th & Constitution, NW (20230) Tel: 202/377-3181
FLORIDA	
Miami	Suite 224, Federal Building 51 SW First Ave. (33130) Tel: 305/536-5267
+Clearwater	128 North Osceola Ave. (34615) Tel: 813/461-0011
+Jacksonville	3100 University Blvd., South Suite 200A (32216) Tel: 904/791-1608
+Orlando	111 North Orange Ave. Suite 1439 (32802) Tel: 407/648-1608
+Tallahassee	Collins Bldg., Room 401 107 West Gaines Street (32304) Tel: 904/448-6469
GEORGIA	
Atlanta	Suite 504, 1365 Peachtree St. (30309) Tel: 404/347-7000
Savannah	120 Barnard St., A-107 (31401) Tel: 912/944-4204
HAWAII	
Honolulu	4106 Federal Bldg. P.O. Box 50026 300 Ala Moana Blvd. (96850) Tel: 808/541-1782
IDAHO	
+Boise	Hall of Mirrors Bldg. (83720) 700 West State St., 2nd Floor Tel: 208/334-3857

ILLINOIS		
Chicago	Room 1406, Mid Continental Plaza Bldg. 55 East Monroe St., (60603) Tel: 312/353-4450	
+Palatine	W. R. Harper College Algonquin & Roselle Rd. (60067) Tel: 312/397-3000; ext. 2532	
+Rockford	515 North Court St. P.O. Box 1747 (61110-2047) Tel: 815/987-8123	
INDIANA		
Indianapolis	1 North Capital, Suite 520 (46204) Tel: 317/226-6214	
IOWA		
Des Moines	817 Federal Bldg. 210 Walnut St. (50309) Tel: 515/284-4222	
KANSAS		
+Wichita	7591 River Park Pl., Suite 580 727 North Waco (67203)	
KENTUCKY		
Louisville	Room 636B, Gene Snyder Courthouse and Customs Bldg./601 W. Broadway (40202) Tel: 502/582-5066	
LOUISIANA		
New Orleans	432 World Trade Center #2 Canal St. (70130) Tel: 504/589-6546	
MAINE		
+Augusta	77 Sewell St. (04330) Tel: 207/622-8249	
MARYLAND		
Baltimore	413 U.S. Customshouse (21202) 40 South Gay and Lombard St. Tel: 301/962-3560	
MASSACHUSETTS		
Boston	World Trade Center, Suite 207 Commonwealth Pier Area (02210) Tel: 617/565-8563	

MICHIGAN		
Detroit	1140 McNamara Bldg. 477 Michigan Ave. Tel: 313/226-3650	(48226)
+Grand Rapids	300 Monroe NW, Room 409 Tel: 616/456-2411	(49503)
MINNESOTA		
Minneapolis	108 Federal Bldg. 110 S. 4th Street	(55401)
MISSISSIPPI		
Jackson	328 Jackson Mall Office Center 300 Woodrow Wilson Blvd. Tel: 601/965-4388	(39213)
MISSOURI		
*St. Louis	7911 Forsyth Blvd., Suite 610 Tel:314/425-3302	(63105)
Kansas City	Room 635, 601 East 12th St. Tel: 816/426-3141	(64106)
MONTANA	Serviced by the Denver District Office	
NEBRASKA		
Omaha	11133 "O" Street Tel: 402/221-3664	(68137)
NEVADA		
Reno	1755 East Plumb Ln., #152 Tel: 702/785-5203	(89502)
NEW HAMPSHIRE	Serviced by Boston District Office	
NEW JERSEY		
Trenton	3131 Princeton Pike Bldg, #6, Suite 100 Tel: 609/989-2100	(08648)
NEW MEXICO		
Albuquerque	5000 Marble NE, Suite 320 Tel: 505/262-6024	(87110)
Santa Fe	c/o Economic Development and Tourism Dept. 1100 St Francis Dr. Tel: 505/827-0264	(87503)

NEW YORK			
	Buffalo	1312 Federal Bldg. 111 West Huron St. Tel: 716/846-4191	(14202)
	+Rochester	121 East Ave. Tel: 716/263-6480	(14604)
	New York	Federal Office Bldg. 26 Fed Plaza, Room 3718 Foley Square Tel: 212/264-0634	(10278)
NORTH CAROLINA			
	Greensboro	324 W. Market St. P.O. Box 1950 Tel: 919/333-5345	(27402)
NORTH DAKOTA		Serviced by Omaha District Office	
OHIO			
	*Cincinnati	9504 Federal Office Bldg. 550 Main St. Tel: 513/684-2944	(45202)
	Cleveland	Room 600, 668 Euclid Ave. Tel: 216/522-4750	(44114)
OKLAHOMA			
	Oklahoma City	5 Broadway Executive Park, Suite 200 6601 Broadway Extension Tel: 405/231-5302	(73116)
	+Tulsa	440 Sam Houston St. Tel: 918/581-7650	(74127)
OREGON			
	Portland	Room 618, 1220 SW 3rd Ave. Tel: 503/221-3001	(97204)
PENNSYLVANIA			
	Philadelphia	475 Allendale Rd, Suite 202 King of Prussia, PA Tel: 215/962-4980	(19406)
Pittsburgh		2002 Federal Bldg., 1000 Liberty Ave. Tel: 412/644-2850	(15222)

PUERTO RICO		
San Juan (Hato Rey)	Room G-55, Federal Bldg. Tel: 809/766-555	(00918)
RHODE ISLAND		
+Providence	7 Jackson Walkway Tel: 401/528-5104	(02903)
SOUTH CAROLINA		
Columbia	Strom Thurmond Federal Bldg. Suite 172 1835 Assembly St. Tel: 803/765-5345	(29201)
+Charleston	J.C. Long Bldg., Room 128 9 Liberty St. Tel: 803/724-4361	(29424)
SOUTH DAKOTA	Serviced by Omaha District Office	
TENNESSEE		
Nashville	Suite 1114, Parkway Towers 404 James Robertson Pkwy Tel: 615/736-5161	(37219-1505)
+Memphis	The Falls Bldg., Suite 200 22 Front St. Tel: 901/521-4137	(38103)
TEXAS		
*Dallas	1100 Commerce St. Room 7A5 Tel: 214/767-0542	(75242-0542)
+Austin	P.O. Box 12724 816 Congress Ave.,Suite 1200	(78711)
Houston	2625 Federal Courthouse 515 Rusk St. Tel: 713/229-2578	(77002)
UTAH		
Salt Lake City	Room 340 U.S. Courthouse, 350 Main St. Tel: 801/524-5116	(84101)
VERMONT	Serviced by Boston District Office	
VIRGINIA		
Richmond	8010 Federal Bldg. 400 N. 8th St. Tel: 804/771-2246	(23240)

WASHINGTON			
	Seattle	3131 Elliot Ave., Suite 300 Tel: 206/442-5616	(98121)
	+Spokane	West 808, West Spokane Blvd., Room 650 Tel: 509/353-2922	(99201)
WEST VIRGINIA			
	Charleston	3402 Federal Bldg. 500 Quarter St. Tel: 304/347-5123	(25301)
WISCONSIN			
	Milwaukee	Federal Bldg. U.S. Courthouse, Room 606 517 East Wise Ave.	(53202)
WYOMING		Service by Denver District Office	

========

NOTE:
- * denotes Regional Office with supervisory regional responsibilities.
- \+ denotes trade specialist at a branch office.

Source: Business America, monthly publication ITA (International Trade Administration).

INTERNATIONAL TRADE ADMINISTRATION
OFFICE OF PUBLICATIONS

Office of Publications
International Trade Administration
Department of Commerce
14th & Constitution, NW
Washington, DC 20230
(202) 377-0632

COMMERCE BUSINESS DAILY
Commerce Business Daily (CBD) is a daily list of government procurement invitations, contract awards, subcontracting leads, sales of surplus property and foreign business opportunities, as well as certain foreign government procurement. The annual subscription cost is $261.00 for first class mail and $208.00 second-class. A six-month trial subscription costs $130.00 for first-class and $104.00 second-class. CBD is available on-line from a sizable list of contractors named in the lower left-hand corner of the front page of each edition. Contact ITA for additional information, or GPO to order a subscription.

EXPORT ENFORCEMENT

Office of Export Enforcement
Bureau of Export Administration
Department of Commerce
14th & Constitution, NW, Room 4616
Washington, DC 20230
(202) 377-8252

This office investigates suspected export control violations and conducts criminal and administrative investigations in the United States. The office conducts pre-license and post-shipment checks overseas and provides regional export enforcement support to U.S. embassies. OEE also coordinates with the U.S. Customs Service on investigative activities and seizure of goods.

EXPORT LICENSE APPLICATION INFORMATION

Office of Export Licensing
Bureau of Export Administration
Department of Commerce
14th & Constitution, NW, Room 1099D
Washington, DC 20230
(202) 377-4811

Export License Application and Information Network (ELAIN) offers U.S. exporters a fast and convenient way to submit license applications to commerce. This time-saving service accepts export license applications electronically for all western country destinations. On receipt, the Commerce Department will review, process, and issue the license electronically for all commodities except supercomputers.

EXPORT LICENSE TRACKING

Office of Export Licensing (OEL)
Bureau of Export Administration
Department of Commerce
14th & Constitution, NW
Washington, DC 20230
(202) 377-2753

The System for Tracking Export License Applications (STELA) answers the question most frequently asked by business: "What is the status of my license application?" Call STELA for a brief recorded message with up-to-the-minute status report on your case. STELA's automated voice response system can be accessed using a push-button phone. STELA can also give exporters authority to ship their goods for those applications approved without conditions. For information on how to use STELA, or if you want to know your case number, call the number listed above.

EXPORT LICENSING

Bureau of Export Administration
Department of Commerce
14th & Constitution, NW
Washington, DC 20230
(202) 377-8536

This office processes all export license applications and issues all export licenses for commodities and technical data controlled for national security, foreign policy, and short supply reasons. The office also administers the international import certificate and delivery verification system. OEL provides assistance to the business community, directs the bureaus' Export Seminar Program, and works with foreign licensing officer counterparts to harmonize licensing procedures.

EXPORT LICENSING NEWSLETTER

Office of Export Licensing
Bureau of Export Administration
Department of Commerce
14th & Constitution, NW, Room 1608
Washington, DC 20230
(202) 377-8731

Office of Export Licensing "Insider" is a free quarterly newsletter published by the Office of Export Licensing, which contains information about OEL, licensing procedures, publications, and seminars, as well as a Q&A section concerning a specific export licensing area.

EXPORT NOW

Public Affairs Office
International Trade Administration
U.S. Department of Commerce
14th & Constitution, NW
Washington, DC 20230
(202) 377-3808

With the dollar substantially declining against the currencies of U.S. major trading partners, the government's efforts to open world markets to U.S. goods has paid off. As American products improved in quality and became more price-competitive abroad, exports became increasingly important in the U.S. economy. Recognizing that this economic climate was ideal for improving the position of U.S. companies in the world markets, then-president Ronald Reagan launched the **EXPORT NOW** program. This major public and private sector cooperative campaign continues to encourage U.S. business firms to take advantage of a window of opportunity that has opened for them to reach fresh markets, increase profits, expand product lines, and create new jobs for Americans. **EXPORT NOW** especially focuses attention on the export opportunities and resources awaiting small- and medium-size companies. Contact this ITA office for more information and/or participation in **EXPORT NOW.**

EXPORT SEMINARS AND WORKSHOPS

U.S. and Foreign Commercial Service
International Trade Administration
U.S. Department of Commerce
14th & Constitution, NW, Room H2106
Washington, DC 20230
(202) 377-5777

Trade specialists can keep you posted on upcoming export workshops, conferences, and seminars in your area, as well as on domestic and overseas trade events that offer good potential for promoting your products or services. Contact this office for more information on export seminars and workshops.

EXPORTER ASSISTANCE

Exporter Assistance Staff
Office of Export Licensing
Bureau of Export Administration
Department of Commerce
14th & Constitution, NW, Room 1099D
Washington, DC 20230
(202) 377-4811

This office's experts can help solve problems or answer most of your questions about how to apply for an export license. By telephone or mail these experts can help you prepare license applications and guide you through export regulations.

Business, industry, and government will always get fast and courteous assistance to help them comply with U.S. Export Control Laws. "The Export License," a step-by-step guide to filling out the export license application, is available free of charge.

EXPORTER SEMINARS
Export Seminar Staff
Bureau of Export Administration
Department of Commerce
14th & Constitution, NW, Room 1608
Washington, DC 20230
(202) 377-8731

One way that the Bureau of Export Administration distributes information necessary to businesses on the export regulations is in the form of Exporter Seminars. These seminars are held in cities all across the country, one each month, as well as an annual update conference in Washington, DC. Seminar instructors are experienced export licensing, policy, and enforcement officials. In addition to conducting training sessions, they meet privately with individual attendees to answer questions and discuss specific export licensing problems. A three-day comprehensive seminar teaches the exporter how to use and interpret the Export Administration Regulations, fill out export license applications, and use the Commodity Control List. The program also updates exporters on export legislation and regulation and provides detailed coverage of distribution licensing and internal control programs, GTE, and LLR licensing, temporary exports, parts and components requirements, technical data transfers, and more. Also conducted are overseas seminars on export licensing for importers and exporters of U.S. commodities. Contact this office for a schedule of seminars and further information.

FEES FOR US&FCS SERVICES

U.S. and Foreign Commercial Service
International Trade Administration
U.S. Department of Commerce
14th & Constitution, NW, Room H2106
Washington, DC 20230
(202) 377-5777

Customs services (such as Comparison Shopping, Commercial News USA, CIMS, and participation in trade show events and other activities) carry participation fees, which are kept as low as possible so that smaller businesses can participate. Counseling services are free of charge. Contact this office for more information on fees.

FOREIGN BUYER PROGRAM

U.S. and Foreign Commercial Service
International Trade Administration
U.S. Department of Commerce
14th & Constitution, NW, Room H2106
Washington, DC 20230
(202) 377-5777

An excellent method for exporters to contact prime businesses is through the US&FCS Foreign Buyer Program. If you have no time to travel overseas, you can participate in an ITA-supported trade show here in the U.S. which is the next best way to meet qualified foreign customers. The US&FCS promotes these shows worldwide to attract foreign buyer delegations, manages an International Business Center, and counsels participating U.S. firms. Contact this ITA office for more information.

FOREIGN EXPORT POLICY

Office of Technology and Policy Analysis (OTPA)
Bureau of Export Administration
Department of Commerce
14th & Constitution, NW
Washington, DC 20230
(202) 377-4188

This office develops and implements overall policies for the licensing of exports. It provides policy guidance and technical expertise for a wide range of technologies and country-specific issues. OTPA is responsible for technical support in the review of both the U.S. and the International Control Lists and for integration of the Military Critical Technologies List.

FOREIGN MARKET RESEARCH

U.S. and Foreign Commercial Service
International Trade Administration
U.S. Department of Commerce
14th & Constitution, NW, Room H2106
Washington, DC 20230
(202) 377-5777

If you want to know how your industry's products are selling around the world, which markets are growing the fastest, which ones are changing, or which ones want to buy more of your products, the

U.S. and Foreign Commercial Service (US&FCS) has the answers for a large number of U.S. industries. US&FCS can provide highly specific analysis of your target markets, tailored to your industry and potential customers. Experienced research analysts continually collect and analyze business and economic data overseas. They search for specific export opportunities and identify market trends affecting trade, investment, and other issues of concern to U.S. businesses interested in trading abroad. The US&FCS develops extensive research on a wide variety of overseas markets. Concise, customized reports can be provided for an industry or country, in print or on diskettes. This wealth of international marketing data is available to you through your nearest US&FCS office. For more information on foreign market research, contact this office.

INTERNATIONAL ECONOMIC POLICY

International Economic Policy
International Trade Commission
U.S. Department of Commerce
14th & Constitution, NW
Washington, DC 20230
(202) 377-3022

The International Economic Policy (IEP) sector, seeking to increase U.S. trade and investment, identifies foreign barriers to commerce and takes part in negotiations to remove them. It also searches for U.S. commercial opportunities overseas and counsels U.S. businesses. Some recent activities of IEP include the following: prepared U.S. positions for the U.S. - Soviet Joint Commercial Commission meeting in Moscow and supported the Secretary of Commerce's participation in its discussions; participated in negotiations that led to the U.S.-Canada Free Trade Area Agreement; conducted seminars throughout the U.S.

MAPS - CIA WORLD AND COUNTRY MAPS

National Technical Information Service
U.S. Department of Commerce
5285 Port Royal Road
Springfield, VA 22151
(703) 487-4650

Hundreds of maps generated by the Central Intelligence Agency are sold through NTIS. There are country maps as well as maps of continents available. Also included are smaller geographical areas and city maps such as Moscow and Vicinity; Middle East Area Oil Fields and Facilities; Israeli Settlement in the Gaza Strip; South Africa: Industrial Activity and Production; Africa Ethnolinguistic Groups; and street maps for Moscow, Shanghai and many other cities.

MATCHMAKERS/TRADE MISSIONS

U.S. and Foreign Commercial Service
International Trade Administration
U.S. Department of Commerce
14th & Constitution, NW, Room H2106
Washington, DC 20230
(202) 377-5777

Trade events can be a profitable part of your marketing plan. You can invest as much money and effort as you choose, depending on your strategy. Trade events can be as simple as sending your product catalogs overseas or as intense as participating in a major international exhibition. You can meet potential customers in person, or send your catalog "on tour" via the "Matchmaker" delegation or trade mission to target markets selected for their high potential for your particular industry. US&FCS will accompany you overseas, arrange a full schedule of customers, and provide complete logistic and promotional support. You will meet senior industry leaders and government officials who can make on-the-spot decisions. US&FCS also sponsors state- and industry-organized trade missions. Contact this office for more information on the "Matchmakers" and the Trade Mission programs.

OVERSEAS TRADE SPECIALISTS

U.S. and Foreign Commercial Service
International Trade Administration
U.S. Department of Commerce
14th & Constitution, NW, Room H2106
Washington, DC 20230
(202) 377-5777

Because US&FCS commercial officers understand firsthand the problems that U.S. companies encounter in their efforts to trade abroad, they are is a strong position to provide a range of services to help companies sell overseas. Trade specialists do on-the-spot market research overseas, including customized individual studies; search for sales leads and qualified agents and distributors for your product; make appointments with key buyers and government officials; and counsel firms frustrated by trade barriers. They speak the host-country language, understand local customs and traditions, as well as trade regulations, and can be a particularly valuable resource when you're visiting overseas markets in person. Contact this office to discuss your marketing needs with an overseas trade specialist.

PUBLIC AFFAIRS OFFICE
Office of Public Affairs
U.S. Department of Commerce
International Trade Administration
14th & Constitution, NW, Room H3414
Washington, DC 20230
(202) 377-3808

Assessments of U.S. Industries
Competitive assessments of U.S. industries are studies of the competitiveness of selected U.S. industries as compared with their foreign counterparts. These assessments are prepared by International Trade Administration industry analysts with the assistance in particular cases of other U.S. government agencies, industry associations, and various private sector groups and individuals. Assessments of more than 45 industries or industry sectors have been published as of June 1989. For an up-to-date listing of each study, contact this office.

SERVICE SECTOR MARKETS OVERSEAS

Services Unit
International Trade Administration
U.S. Department of Commerce
14th & Constitution, NW
Washington, DC 20230
(202) 377-1461

Because the U.S. is very competitive in the services sector, the Services Unit plays an important role in trade promotion efforts. This office provides information on the operation of services sectors in other countries and on specific trade opportunities. Contact this office for more information.

SERVICES FOR EXPORTERS

U.S. and Foreign Commercial Service
International Trade Administration
U.S. Department of Commerce
14th & Constitution, NW, Room H2106
Washington, DC 20230
(202) 377-5777

U.S. companies are strategically positioned to profit from exporting their products and services to many nations, but selling goods and services in foreign markets doesn't have to be complicated and confusing. Firsthand knowledge of overseas markets and the exporting process can greatly increase your company's chance of exporting success. However, finding the information and advice you need to enter--and succeed--in world markets can be a major problem. So, the U.S. Foreign Commercial Service has trade professionals around the world dedicated to your export success. Experts are available to help you export your goods and services in virtually all major world markets, and they offer highly effective services to help you every step of the way. For more information on services for exporters, please contact this office.

STARTING POINTS FOR OVERSEAS TRADING

U.S. and Foreign Commercial Service
International Trade Administration
U.S. Department of Commerce
14th & Constitution, NW, Room HCHB1066
Washington, DC 20230
(202) 377-3181

The first stop a U.S. company should make to take advantage of government export programs is the nearest Commerce Department (ITA) district office. Through the Commerce Department, including practical information about overseas market opportunities, developed by commercial officers abroad and by industry and country desk officers in Washington, D.C. The district offices can also direct companies toward other government and private sector export services. Each office is headed by a director and supported by trade specialists and other staff. These professionals help company decision-makers gain a basic understanding of profitable opportunities in exporting and assist them in evaluating the company's market potential overseas. To find out how the Foreign Commercial Service's programs can benefit your company, call your nearest U.S. Department of Commerce District Office.

Most of the listed offices maintain extensive business libraries containing the Commerce Department's latest reports.

FOREIGN EXCHANGE AND GOLD OPERATIONS

Foreign Exchange and Gold Operations
U.S. Department of the Treasury
1500 Pennsylvania Ave., NW, Room 5037
Washington, DC 20220
(202) 566-2773

This office monitors the foreign money markets for the Department of the Treasury. Contact them for more information.

FOREIGN TRADE AND INVESTMENT POLICY

Deputy Asst. Sec. for Trade and Investment Policy
Office of the Asst. Sec. for International Affairs
U.S. Department of the Treasury
1500 Pennsylvania Ave., NW
Washington, DC 20220
(202) 566-2748

This office formulates trade and investment policies. Responsibilities in this area also include international trade, international investment, East-West Policy, trade finance, and commodity policy.

IMPORTATION UNDER BOND

Public Information Office
U.S. Customs Service
U.S. Department of the Treasury
P.O. Box 7407
Washington, DC 20044
(202) 566-8195

Temporary importation under bond (TIB) is a procedure whereby, under certain conditions, merchandise may be entered into the Customs territory temporarily free of duty by posting a bond. In the bond, the importer agrees to export or destroy the merchandise within a specified time or pay liquidated damages, generally equal to twice the normal duty. The booklet, "Temporary Importation Under Bond," explains the bond exportation requirements and the classes of goods involved.

INTERNATIONAL CAPITAL BANKING STATISTICS

Office of Data Management
Office of the Assistant Secretary for Intn'l Affairs
U.S. Department of the Treasury
1500 Pennsylvania Ave., NW, Room 5127
Washington, DC 20220
(202) 566-5473

Treasury International Capital Forms are filed for the purpose of gathering timely information on international capital movements. Banks, other depository institutions, international banking facilities, bank holding companies, brokers, and dealers in the United States must file reports. The organizations describe liabilities to, or claims on, foreigners, whether they are individuals or institutions. The Federal Reserve Bank of New York acts as the agent and receives the filed reports. The address is: International Reports Division, International Reports and Support Dept., 4th Fl., Federal Reserve Bank of New York, 33 Liberty Street, New York, NY, 10045.

INTERNATIONAL CAPITAL NON-BANKING STATISTICS

Office of Data Management
Office of the Assistant Secretary for Int'l Affairs
U.S. Department of the Treasury
1500 Pennsylvania Ave., NW, Room 5127
Washington, DC 20220
(202) 566-5473

International Capital Non-Banking Forms provide information on the liability and claim positions of the United States with respect to foreign countries. This aids in determining the financial position of the United States and the movement of capital between the United States and foreign countries. Unlike the International Capital Banking Forms, these forms are filed by exporters, importers, industrial commercial concerns, non-banking financial institutions, and foreign sales corporations. The Federal Reserve Bank of New York acts as the agent for the collection of these forms. The address is: International Reports Division, Federal Reserve Bank of New York, 33 Liberty Street, New York, NY 10045.

INTERNATIONAL COMMODITIES

Office of Commodity Policy
Office of the Assistant Secretary for Int'l Affairs
U.S. Department of the Treasury
1500 Pennsylvania Ave., NW, Room 5423
Washington, DC 20220
(202) 566-2561

This office monitors international commodity developments of such products as wheat and meat imports, tropical timber, iron ore, copper, cocoa, coffee, and sugar. They also stay abreast of activity involving the International Rubber Agreement, the United Nations Conference of Trade and Development Commodities Issues, oceans policy, the world oil market, and international energy policy.

FOREIGN COMPETITION

The Office of Investigations
U.S. International Trade Commission
500 E. Street, SW, Room 615
Washington, DC 20436
(202) 252-1161

The ITC investigates issues relating to U.S. and foreign customs laws; the volume of importation in comparison with domestic production and consumption; the conditions, causes, and effects relating to competition of foreign industries with those of the United States; and all other factors affecting competition between U.S. and imported articles.

FREE-TRADE AGREEMENTS

The Trade Report Division
Office of Economics
U.S. International Trade Commission
500 E. Street, SW, Room 602
Washington, DC 20436
(202) 252-2126

The ITC conducts some very complex, sophisticated studies regarding trade. One of these investigations, for example, examines the pros and cons of initiating negotiations with Japan to explore the possibility of the United States-Japan free-trade agreement. The ITC is also conducting a similar study involving possible free-trade agreements with Taiwan, the Republic of Korea, the Association of Southeast Asian Nations, and the entire Pacific Rim. For more information concerning these studies and other free-trade issues, contact this office.

GLOBAL COMPETITIVENESS OF SELECTED INDUSTRIES

Office of the Secretary
U.S. International Trade Commission
500 E. Street, SW
Washington, DC 20436
(202) 252-1000

To assess the global competitiveness of selected industries, the ITC has conducted studies which cover textiles, steel, automotive parts, optical fibers, and oilseeds. Contact this office for more information concerning these studies and other issues of competition.

HISTORICAL TRADE INFORMATION

Office of the Secretary
U.S. International Trade Commission
500 E. Street, SW
Washington, DC 20436
(202) 252-1000

This office has been a depository for historical information on the Commission since 1916. Contact this office for information on its holdings.

IMPORTED ARTICLES SUBSIDIZED OR SOLD LESS THAN FAIR VALUE

Office of Investigations
U.S. International Trade Commission
500 E. Street, SW, Room 615
Washington, DC 20436
(202) 252-1161

The ITC conducts investigations to determine whether there is reasonable indication of material injury to, threat of material injury to, or the establishment of, and industry in the U.S. by reason of imports of foreign merchandise allegedly being subsidized or sold at less than fair value.

For information on current and past investigations, contact this office.

INTERNATIONAL COMPETITIVENESS

Office of Industries
U.S. International Trade Commission
500 E. Street, SW, Room 615
Washington, DC 20436
(202) 252-1296

This office studies the international competitiveness of U.S. producers in such areas as textile mill products, oilseeds, steel sheet and strips, auto parts, optical fibers, certain vegetables, and steels. For more information on these and other studies, contact this office.

INTERNATIONAL ECONOMIC REVIEW

Trade Reports Division
Office of Economics
U.S. International Trade Commission
500 E. Street, SW, Room 602
Washington, DC 20436
(202) 252-1216

The "International Economic Review," a monthly review of trade and policy developments, provides information on significant economic and policy developments abroad that may affect U.S. commerce. The review is distributed to members of Congress and the public on request.

INTERNATIONAL TRADE STUDIES

Office of the Secretary
U.S. International Trade Commission
500 E. Street, SW
Washington, DC 20436
(202) 252-1216

Following requests from the U.S. president, the House Ways and Means Committee, the Senate Finance Committee, either branch of Congress, or on its own motion, the ITC conducts studies, investigations, and research projects on a broad range of topics relating to international trade. Public reports of these studies, investigations, and research projects are issued in most cases. The commission also keeps informed of the operation and effect of provisions relating to duties or other import restrictions of the United States contained in various trade agreements. Occasionally, the commission is required by statute to perform specific trade-related studies.

RESEARCH METHODS

Investigation Support Division
Office of Economics
U.S. International Trade Commission
500 E. Street, SW, Room 602
Washington, DC 20436
(202) 252-1216

ITC staff use economic models to determine the effects of certain tariffs and quotas on imports. In addition to providing estimates of supply and characteristics of demand in all final investigations, these ITC economists provide estimates to outside parties to investigate prior to commission hearings so that these parties are able to review and comment on the estimates. In final investigations, staff economists also include estimates of the effect of dumping on the price and volume of sales of domestic products. Estimates are based on methodology developed in this office. Contact this office for more information on tariffs and import quotas.

EXPORT INFORMATION SYSTEM (XIS)

Small Business Answer Desk
U.S. Small Business Administration
1725 Eye St., NW, Room 414
Washington, DC 20416
(202) 653-7561
(800) 368-5855

The Small Business Administration's Export Information System provides an initial screening of world markets to determine the export potential of a commodity or manufactured product and contains data on exports and imports supplied by members of the United Nations. SBA offices use the XIS to assist small firms in finding markets for their product overseas, in evaluating trends in those markets, and in assessing the types of competition in specific country markets. Contact your SBA office for additional information.

EXPORT LEGAL ADVICE NETWORK (ELAN)

Small Business Answer Desk
U.S. Small Business Administration
1725 Eye St., NW, Room 414
Washington, DC 20416
(202) 653-7561
(800) 368-5855

The Export Legal Assistance Network (ELAN) is a cooperative program between SBA and the Federal Bar Association. ELAN provides free, one-time legal consultations to new-to-export small companies. The FBA maintains rosters of attorneys in 40 cities who are knowledgeable in export-related legal matters and available to assist small firms. For more information on ELAN, contact the SBA office in your area.

EXPORT REVOLVING LINE OF CREDIT LOAN (ERLC)

Small Business Answer Desk
U.S. Small Business Administration
1725 Eye St., NW, Room 414
Washington, DC 20416
(202) 653-7561
(800) 368-5855

To assist small businesses in exporting their products and services abroad, SBA has established the Export Revolving Line of Credit Loan Program (ERLC). Any number of withdrawals and repayments can be made as long as the dollar limit of the line is not exceeded and the disbursements are made within the stated maturity period. For more information on eligibility, use of proceeds, amount of loan, maturity, interest rates, fees, collateral, special program requirements, and how to apply, contact your SBA office.

EXPORT TRAINING PROGRAMS

Small Business Answer Desk
U.S. Small Business Administration
1725 Eye St., NW, Room 414
Washington, DC 20416
(202) 653-7561
(800) 368-5855

Export training programs are co-sponsored by SBA district offices with the U.S. Department of Commerce and others interested in international trade. Emphasis is placed on the practical application of successful exporting procedures to small business. Contact your SBA office for more information.

INTERNATIONAL TRADE

Assistance and Publications
Office of International Trade
U.S. Small Business Administration
1441 L St., NW, Room 501A
Washington, DC 20416
(202) 653-7794

This office provides information and assistance to small business owners completely new to exporting and those already exporting but planning to expand their markets. OIT works closely with SBA field offices in providing a range of outreach and educational programs. Contact this office for more information or to obtain a free copy of their publication, "The Exporter's Guide to Federal Resources for Small Business." Another publication issued from this office, "Market Overseas with U.S. Government Help", is available for $1.00 by writing SBA, P.O. Box 15434, Fort Worth, TX, 76119.

INTERNATIONAL TRADE LOANS

Small Business Answer Desk
U.S. Small Business Administration
1725 Eye St., NW, Room 414
Washington, DC 20416
(202) 653-7561
(800) 368-5855

The International Trade Loan program has been established to help more small businesses both compete more effectively with import competition, as well as significantly expand existing export markets. For more information on these loans, including the amount of loan, percentage of guaranty, eligibility, collateral, and maturity, contact the SBA office in your area.

MATCHMAKER TRADE MISSIONS

Office of International Trade
U.S. Small Business Administration
1441 L St., NW, Room 501A
Washington, DC 20416
(202) 653-7794

Cosponsored by SBA and the U.S. Department of Commerce, this program arranges direct contact for American firms with potential partners in new international markets. Contact the office above or your SBA office for more information.

CERTIFICATION FOR DOCUMENTS/TRAVEL ABROAD

Authentications
Foreign Affairs Center
Bureau of Administration
Department of State
2201 C St., NW, Room 2815
Washington, DC 20520
(202)647-7735

Several steps are necessary in order to get a document authenticated for use by a foreign government. For legal papers to do business in a foreign government (such as a power of attorney), the papers must be notarized, signed by the Clerk of Court, and then sent to the Secretary of State for the State Seal. This office authenticates the State Seal, verifying that it is a legal document. Foreign governments often require that any legal document, whether personal, educational or business related, be authenticated. This includes birth certificates, marriage licenses, divorce papers, or school transcripts.

CITIZENS ARRESTED OVERSEAS

Citizens Emergency Center
Overseas Citizens Service
Bureau of Consular Affairs
U.S. Department of State
2201 C St., NW, Room 4800
Washington, DC 20520
(202)647-5225

The Citizens Emergency Center monitors the cases of Americans arrested abroad and acts as a liaison between the prisoner's family and counselor officers overseas. A consular officer visits the

American as soon as possible, provides information regarding the foreign legal system and a list of attorneys, and offers other assistance such as contacting family and friends. The consular officer's role in arrest cases is one of observation and support, regularly visiting the prisoner and checking his or her welfare, and monitoring human rights and the status of the case. The center assists in transferring of funds. Also, when a prisoner's health or life is endangered by inadequate diet or medical care provided by the local prison, dietary food supplements and/or medical care may be arranged through a U.S. government loan authorized under the Emergency Medical and Dietary Assistance Program.

CITIZENS EMERGENCY CENTER

Citizens Emergency Center
Overseas Citizens Service
Bureau of Consular Affairs
U.S. Department of State
2201 C St., NW, Room 4800
Washington, DC 20520
(202)647-5225

This center provides assistance to U.S. citizens and protects their welfare abroad in the following ways:

Arrests - See Citizens Arrested Overseas

Financial Assistance - The center assists Americans overseas who find themselves in financial trouble. They first attempt to locate private sources of funds from family or friends, and then assist with the transfer of funds to the individual. If none can be found, the center will approve a repatriation loan which will pay for the individual's direct return to the nearest port of entry in the U.S.

Medical Assistance - The center assists with handling the problems of Americans who become physically or mentally ill while traveling or living abroad. The center also locates and notifies family or friends and transmits private funds. When necessary, the staff will assist in the return of the ill or injured person to the U.S. with appropriate medical escort. Full expenses must be borne by family members.

Deaths - The consular officer reports the death of a U.S. citizen to the next of kin and will assist in making arrangements for local burial or for return of the body to the U.S. Costs must be borne by family members.

Welfare/Whereabouts - The center relays the request for assistance and all pertinent data available on the missing individual to the U.S. Embassy or consulate responsible for the area where the individual is believed to be traveling or residing, and the consulate officer then attempts to locate this individual. In cases of disasters such as earthquakes or plane crashes, the center ascertains the names of U.S. citizens involved and informs their families.

Travel Advisories - The center gives advice to the public and U.S. Foreign Service posts on the advisability of travel to certain countries or areas.

Search and Rescue - The center monitors the search and rescue efforts outside the U.S., such as attempting to locate missing planes or boats that might be carrying Americans.

EAST - WEST TRADE POLICY

Bureau of Economic and Business Affairs
Department of State
2001 C Street, NW, Room 3819
Washington, DC 20520
(202) 647-2875

This office, as one of two offices in International Trade Controls, has broad responsibilities for the conduct of U.S. economic policy, including policies between the U.S. and the USSR. It is primarily responsible, however, for the negotiation of strategic trade agreements with newly industrialized nations, along with being charged with the administration of worldwide foreign policy export controls. Information is available on importing or exporting products, as well as on doing business with foreign countries.

EXECUTIVE - DIPLOMAT SEMINARS

Office of Public Programs
Department of State
2001 C Street, NW, Room 3819
Washington, DC 20520
(202) 647-2875

The State Department holds seminars designed for corporate vice-presidents who do business abroad. These two-day seminars, offered twice yearly, begin with a discussion of global foreign policy objectives and then focus on economic topics and business opportunities. Contact this office for information on scheduling.

EXPORT CONTROL - STRATEGIC

Dual Use Technology

Office of International Trade Control
Bureau of Economic and Business Affairs
Department of State
2201 C. Street, NW, Room 3529
Washington, DC 20520
(202) 647-1625

This office provides foreign policy guidance pertaining to U.S. export controls for national security and foreign policy purposes and for non-strategic trade with Eastern Europe, the Soviet Union, the People's Republic of China, and other non-market economy countries. It represents the Department of State in advising the Department of Commerce on U.S. foreign policy aspects of U.S. strategic export control under the Export Administration Act.

NUCLEAR EXPORT AND IMPORT CONTROL

Nuclear Energy and Energy Technology Affairs
Bureau of Oceans and International Environmental and Scientific Affairs
U.S. Department of State
2201 C. Street, NW, Room 7820
Washington, DC 20520
(202) 647-4701

This office works closely with the Nuclear Regulatory Commission with regard to the licensing requests for nuclear imports and exports. They also study the development and transfer of nuclear technology in foreign countries. This office monitors the Nuclear Non-Proliferation Treaty, making sure that the countries which signed the treaty are following its regulations. They also oversee the export and import of items which are not nuclear but which could be used for nuclear purposes, such as supercomputers. The export of these items must follow the guidelines set in the Nuclear Non-Proliferation Treaty.

AGENCY FOR INTERNATIONAL DEVELOPMENT (AID)

U.S. International Development Cooperation Agency
320 21st Street, N.W.
Washington, DC 20523
202/632-9620 (General Information)
202/235-1720 (AID)

AID finances through loans and grants the purchase of U.S. products and services by public and private sector importers in developing countries. AID financial aid goes to the foreign government, which then allocates the foreign exchange thus provided among importing agencies and firms. Programs and services provided in this area are Small and Disadvantaged Business Utilization and Foreign Trade Opportunities.

U.S. TRADE REPRESENTATIVE

Office of the U.S. Trade Representative
Winder Building
600 17th Street, NW
Washington, DC 20506
(202) 395-3515

This office develops and coordinates U.S. international trade, commodity, and direct investment policy, and leads or directs negotiations with other countries on such matters. There are two offices, one in Washington, DC, and the other in Geneva, Switzerland. The U.S. Trade Representative is a Cabinet-level official with the rank of Ambassador, who acts as the principal trade advisor, negotiator, and spokesperson for the President on trade and related investment matters.

6.4.2 Associations

6.4.2.1 The Alliance to Save Energy
1725 K Street, N.W., Suite #509
Washington, DC 20006-1401
(202) 857-0666

6.4.2.2 Association of Energy Engineers
4025 Pleasantdale Road, Suite #420
Atlanta, GA 30340
(404) 477-5083

6.4.2.3 Overseas Sales & Marketing Assoc. of America
5715 N. Lincoln Ave.
Chicago, IL 60659
(818) 679-6070

6.4.2.4 Export Management Assoc. of California
14549 Victory Blvd., Suite #5
Van Nuys, CA 91411
(818) 782-3350

6.4.3 Business by Sector and Industry

6.4.3.1 Public

6.4.3.2 Private

6.4.3.3 Telecommunications

6.4.3.3.1 "Finding Their Voice, Telecommunications in Eastern Europe," The Economist, February 8th, 1992.

6.5 Resource and Data Base

6.5.1 Market Information
- 6.5.1.1 Economic Developments
- 6.5.1.2 Legal Developments
- 6.5.1.3 Currency and Foreign Exchange

6.5.2 Trade Leads
- 6.5.2.1 Government contact for doing business
- 6.5.2.2 Associations
- 6.5.2.3 Commercial
- 6.5.2.4 News Sources - Eastern Europe and Baltics

6.5.3 Contact Lists by Industry
- 6.5.3.1 Government
- 6.5.3.2 Associations
- 6.5.3.3 Commercial

  Central & Eastern European Business Information: List of Reports Covering Trade, Markets, Finance, Investment and Joint Venture Leads

APPENDIX A - DEMOGRAPHICS: POLAND

This appendix was made available by the U.S. Department of Commerce, International Trade Administration, Washington, DC 20230.

GEOGRAPHY

Total area: 312,680 km: land area 304,510 km^2
Comparative area: slightly smaller than New Mexico
Land boundaries: 2,980 km total: Czechoslovakia 1,309 km, Germany 156 km, USSR 1,215 km
Coastline: 491 km
Maritime claims: *Territorial.sea*: 12 nm
Climate: temperate with cold, cloudy, moderately severe winters with frequent precipitation; mild summers with frequent showers and thundershowers
Terrain: mostly flat plain, mountains along southern border
Natural resources: coal, sulfur, copper, natural gas, silver, lead, salt
Land use: arable land 46%; permanent crops 1%; meadows and pastures 13%; forest and woodland 28%; other 12%; includes irrigated NEGL%
Environment: plain crossed by a few north-flowing, meandering streams; severe air and water pollution in south
Note: historically an area of conflict because of flat terrain and the lack of natural barriers on the North European Plain

PEOPLE

Population: 37,799,638 (July 1991), growth rate 0.1% (1991)
Birth rate: 14 births/1,000 population (1991)
Death rate: 9 deaths/1,000 population (1991)
Net migration rate: 4 migrants/1,000 population (1991)
Infant mortality rate: 12 deaths/1,000 live births (1991)
Life expectancy at birth: 69 years male, 77 years female (1991)
Total fertility rate: 2.1 children born/woman (1991)
Nationality: noun—Pole(s); adjective—Polish
Ethnic divisions: Polish 97.6%, German 1.3%, Ukrainian 0.6%, Belorussian (Byelorussian) 0.5% (1990 est.)
Religion: Roman Catholic 95% (about 75% practicing), Russian Orthodox, Protestant, and other 5%
Language: Polish
Literacy: 98% (male 99%, female 98%) age 15 and over can read and write (1978)
Labor force: 17,104,000; industry and construction 36.1%; agriculture 27.3%; trade, transportation, and communications 14.8%; government and other 21.8% (1989)
Organized labor: trade union pluralism

ECONOMY

Overview: The economy. except tor the agricultural sector. had followed the Soviet model of state ownership and control of productive assets. About 75% of agricultural production had come from the private sector and the rest from state farms. The economy has presented a picture of moderate but slowing growth against a background of underlying weaknesses in technology and worker motivation. GNP dropped by 2.0% in 1989 and by a further 8.9% in 1990. The inflation rate. after falling sharply from the 1982 peak of 100% to 22% in 1986. rose to a galloping rate of 640% in 1989 and dropped back to 250% in 1990. Shortages of consumer goods and some food items worsened in 1988-89. Agricultural products and coal are among the biggest hard currency earners, but manufactures are increasing in importance. Poland. with its hard currency debt of $48.5 billion. is severely limited in its ability to import much-needed hard currency goods. The sweeping political changes of 1989 disrupt ed normal economic channels and exacerbated shortages. In January 1990, the new Solidarity-led government adopted a cold turkey program for transforming Poland to a market economy. The government moved to eliminate subsidies, free prices. make the zloty convertible. and. in general, halt the hyperinflation. These financial measures were accompanied by plans to privatize the economy in stages. While inflation fell to an annual rate of 77.5% by November of 1990. the rise in unemployment and the drop in living standards have led to growing popular discontent and to a change of government in January 1991. The new government is continuing the previous government's economic program. while trying to speed privatization and to better cushion the populace from the dislocations associated with reform. Substantial outside aid will be needed if Poland is to make a successful transition in the 1990s.

GNP: $158.5 billion, per capita$4,200; real growth rate—8.9% (1990 est.)
Inflation rate (consumer prices): 250% (1990 est.)
Unemployment rate: 6.1% (end-December 1990)
Budget: revenues $20.9 billion; expenditures $23.4 billion, including capital expenditures of $2.8 billion (1989)
Exports: $12.9 billion (f.o.b., 1989); *commodities*—machinery and equipment 38%; fuels, minerals, and metals 21%; manufactured consumer goods 15%; agricultural and forestry products 4% (1989); *partners*—USSR 25%, FRG 14%, UK 6.5%, Czechoslovadia 5.5% (1989)
Imports: $12.8 billion (f.o.b., 1989); *commodities*—machinery and equipment 37%; fuels, minerals, and metals 31%; manufactured consumer goods 17%; agricultural and forestry products 5% (1989); *partners*—USSR 18%, FRG 16%, Austria 6%, Czechoslovadia 6% (1989)
External debt: $48.5 billion rate—23% (State sector 1990 est.)
Electricity: 31,530,000 kW capacity; 136,300 million kWh produced, 3,610 kWh per capita (1990)
Industries: machine building, iron and steel, extractive industries, chemicals, shipbuilding, food processing, glass, beverages, textiles
Agriculture: accounts for 15% of GNP and 27% of labor force; 75% of output from private farms, 25% from state farms; productivity remains low by European standards; leading European producer of rye, rapeseed, and potatoes; wide variety of other crops and livestock; major exporter of pork products; normally self-sufficient in food
Economic aid: donor—bilateral aid to non-Communist less developed countries, $2.2 billion (1954-89)
Currency: zloty (plural—zlotych); 1 sloty (Zl) = 100 groszy
Exchange rates: zlotych (Zl) per US$1—11,100.00 (May 1991), 9,500 (1990), 1,439.18 (1989), 430,55 (1988), 265.08 (1987), 175.29 (1986), 147.14 (1985)
Fiscal year: calendar year

BEST PROSPECTS FOR TRADE AND INVESTMENT

Telecommunications Equipment
Food Processing and Packaging Equipment

Management and Consulting Services
Printing and Graphic Arts Equipment
Pollution Control Equipment
Air Conditioning and Refrigerations Equipment
Household Consumer Goods
Computers and Peripherals
Electrical Power Systems
Hotel and Restaurant Equipment
Furniture
Sporting Goods
Medical Equipment
Industrial Process Controls
Consumer Electronics
Travel and Tourism Services
Automobile Parts
Construction Equipment
Textile Products
Automobile and Light Trucks

APPENDIX B - DEMOGRAPHICS: HUNGARY

This appendix was made available by the U.S. Department of Commerce, International Trade Administration, Washington, DC 20230.

GEOGRAPHY

Total area: 93,030 km^2; land area: 92,340 km^2
Comparative area: slightly smaller than Indiana
Land boundaries: 2,251 km total; Austria 366 km, Czechoslovakia 676 km, Romania 443 km, USSR 135 km, Yugoslavia 631 km
Coastline: none—landlocked
Maritime claims: none—landlocked
Disputes: Nagymaros Dam dispute with Czechoslovakia
Climate: temperate; cold, cloudy, humid winters; warm summers
Terrain: mostly flat to rolling plains
Natural resources: bauxite, coal, natural gas, fertile soils
Land use: arable land 54%; permanent crops 3%; meadows and pastures 14%; forest and woodland 18%; other 11%; includes irrigated 2%
Environment: levees are common along many streams, but flooding occurs almost every year
Note: landlocked; strategic location astride main land routes between Western Europe and Balkan Peninsula as well as between USSR and Mediterranean basin

PEOPLE

Population: 10,558,001 (July 1991), growth rate—0.1% (1991)
Birth rate: 12 births/1,000 population (1991)
Death rate: 13 deaths/1,000 population (1991)
Net migration rate: 0 migrants/1,000 population (1991)
Infant mortality rate: 14 deaths/1,000 population (1991)
Life expectancy at birth: 68 years male, 76 years female (1991)
Total fertility rate: 1.8 children born/woman (1991)
Nationality: noun—Hungarian(s); adjective—Hungarian
Ethnic divisions: Hungarian 96.6%, German 1.6%, Slovak 1.1%, Southern Slav 0.3%, Romanian 0.2%
Religion: Roman Catholic 67.5%, Calvinist 20)%, Lutheran 5.0%, atheist and other 7.5%
Language: Hungarian 98.2%, other 1.8%
Literacy: 99% (male 99%, female 98%) age 15 and over can read and write (1980)
Labor force: 4,860 services, trade, government, and other 43.2%, industry 30.9%, agriculture 18.8%, construction 7.1% (1988)
Organized labor: 96.5% of labor force; Central Council of Hungarian Trade Unions (SZOT) includes 19 affiliated unions, all controlled by the government; independent unions legal; may be as many as 12 small independent unions in operation

ECONOMY

Overview: Agriculture is an important sector, providing sizable export earnings and meeting domestic food needs. Industry accounts for about 40% of GNP and 30% of employment. About 40%

of Hungary's foreign trade is with the USSR and Eastern Europe and a third is with the EX. Low rates of growth reflect the inability of the Soviet-style economy to modernize capital plant and motivate workers. GNP declined by 1% in 1989 and by an estimated 6% in 1990. Since 1985 external debt has more than doubled, to over $20 billion. In recent years Hungary has experimented widely with decentralized and market-oriented enterprises. The newly democratic government has renounced the Soviet economic growth model and plans to open the economy to wider market forces and to much closer economic relations with Western Europe, Prime Minister Antall has declared his intention to move forward on privatization of state enterprises, provision for bankruptcy, land reform, and marketization of international trade, but concerns over acceptable levels of unemployment and inflation may slow the reform process.
GNP: $60.9 billion, per capita $5,800; real growth rate -5.7% (1990 est.)
Inflation rate (consumer prices): 30% (1990 est.)
Unemployment rate: 1.7% (1990)
Budget: revenues $18.2 billion; expenditures $18.3 billion, including capital expenditures of $805 million (1989)
Exports: $10.2 billion (f.o.b. 1989); *commodities*—capital goods 33%, foods 25%, consumer goods 16%, fuels 1.5%, other 24.5%; *partners*—USSR and Eastern Europe 42%, developed countries 37.4%, less developed countries 20.6% (1989)
Imports: $10.1 billion (c.i.f. 1989); *commodities*—capital goods 15%, fuels 20%, manufactured consumer goods 12.4%, agriculture 5%, other 47.6%; *partners*—USSR and Eastern Europe 34.9%, developed countries 45.5%, less developed countries 16.6%, US 3%
External debt: $20.7 billion (1989)
Industrial production: growth rate -7.9% (1990 est.)
Electricity: 7,800,000 kW capacity; 30,400 million kWh produced, 2,870 kWh per capita (1990)
Industries: mining, metallurgy, engineering industries, processed foods, textiles, chemicals (especially pharmaceuticals)
Agriculture: including forestry, accounts for about 15% of GNP and 19% of employment; highly diversified crop-livestock farming; principal crops—wheat, corn, sunflowers, potatoes, sugar beets; livestock—hogs, cattle, poultry, dairy products; self-sufficient in food output
Economic aid: donor—$2.0 billion in bilateral aid to non-Communist less developed countries (1962-89)
Currency: forint (plural—forints); 1 forint (Ft) = 100 filler
Exchange rates: forints (Ft) per US$1—60.95 (December 1990), 63.21 (1990), 59.07 (1989), 50.41 (1988), 46.97 (1987), 45.83 (1986), 50.12 (1985)
Fiscal year: calendar year

BEST PROSPECTS FOR TRADE AND INVESTMENT

Food Processing and Packaging
Telecommunications Equipment
Tourism
Computers and Peripherals
Construction Equipment
Medical Equipment
Pollution Control Equipment
Household Consumer Goods

APPENDIX C - DEMOGRAPHICS: CZECHOSLOVAKIA

This appendix was made available by the U.S. Department of Commerce, International Trade Administration, Washington, DC 20230.

GEOGRAPHY

Total area: 127,870 km^2; land area; 125,460 km^2
Comparative area: slightly larger than New York State
Land boundaries: 3,446 km total; Austria 548 km, Germany 815 km, Hungary 676 km, Poland 1,309 km, USSR 98 km
Coastline: none—landlocked
Maritime claims: none—landlocked
Disputes: Nagymaros Dam dispute with Hungary
Climate: temperate; cool summers; cold, cloudy, humid winters
Terrain: mixture of hills and mountains separated by plains and basins
Natural resource: coal, timber, lignite, uranium, magnesite, iron ore, copper, zinc
Land use: arable land 40%; permanent crops 1%; meadows and pastures 13%; forest and woodland 37%; other 9%; includes irrigated 1%
Environment: infrequent earthquakes; acid rain; water pollution; air pollution
Note: landlocked; strategically located astride some of the oldest and most significant land routes in Europe; Moravian Gate is a traditional military corridor between the North European Plain and the Danube in central Europe

PEOPLE

Population: 15,724,940 (July 1991), growth rate 0.3% (1991)
Birth rate: 14 births/1,000 population (1991)
Death rate: 11 deaths/1,000 population (1991)
Net migration rate: NEGL migrants/1,000 population (1991)
Infant mortality rate: 11 deaths/1,000 live births (1991)
Life expectancy at birth: 69 years male, 77 years female (1991)
Total fertility rate: 1.9 children born/woman (1991)
Nationality: noun—Czechoslovakian(s); adjective—Czechoslovak
Ethnic divisions: Czech 62.9%, Slovak 31.8%, Hungarian 3.8%, Polish 0.5%, German 0.3%, Ukrainian 0.3%, Russian 0.1%, other 0.3%
Religion: Roman Catholic 50%, Protestant 20%, Orthodox 2%, other 28%
Language: Czech and Slovak (official), Hungarian
Literacy: 99% (male NA%, female NA%) age 15 and over can read and write (1970 est.)
Labor force: 8,200,000 (1987); industry 36.9%, agriculture 12.3%, construction, communications, and other 50.8% (1982)
Organized labor: Czech and Slovak Confederation of Trade Unions (CSKOS); new independent trade unions forming

ECONOMY

Overview: Czechoslovakia is highly industrialized and has a well-educated and skilled labor force. Its industry, transport, energy sources, banking, and most other means of production are state owned. The country is deficient, however, in energy and in many raw materials. Moreover, its aging capital plant lags well behind West European standards. Industry contributed over 50% to GNP and construction contributed 10%. About 95% of agricultural land is in collectives or state farms. The centrally planned economy has been tightly linked in trade (80%) to the USSR and Easter Europe. Growth has been sluggish, averaging less than 2% in the period 1982-89. GNP per capita is the highest in Eastern Europe. As in the rest of Eastern Europe, the sweeping political changes of 1989-90 have been disrupting normal channels of supply and compounding the government's economic problems. Having eased restrictions on private enterprise in 1990 and having adjusted some key prices, Czechoslovakia is now implementing a broad two-year program to make the difficult transition from a command to a market economy. Inflation and unemployment are beginning to rise, albeit from comparatively low levels.
GNP: $120.3 billion, per capita $7,700; real growth rate -2.9% (1990 est.)
Inflation rate (consumer prices): 9% (1990)
Unemployment rate: officially 0.8% (1990)
Budget: revenues $17.1 billion; expenditures $16.8 billion, including capital expenditures of $1.5 billion (1991)
Exports: $144.4 billion (f.o.b., 1989); *commodities*—machinery and equipment 42.7%; fuels, minerals, and metals 16.4%; agricultural and forestry products 12.5%, other 28.4%; *partners*—USSR, GDR, Poland, Hungary, FRG, Yugoslavia, Austria, Bulgaria, Romania, US
Imports: $14.3 billion (f.o.b., 1989); *commodities*—machinery and equipment 38.6%; fuels, minerals, and metals 24.1%; agricultural and forestry products 16.4%; other 20.9%; *partners*—USSR, GDR, Poland, Hungary, FRG, Yugoslavia, Austria, Bulgaria, Romania, US
External debt: $7.6 billion, hard currency indebtedness (September 1990)
Industrial production: growth rate -3.3% (1990 est.)
Electricity: 23,000,000 kW capacity; 90,000 million kWh produced, 5,740 kWh per capacity (1990)
Industries: iron and steel, machinery and equipment, cement, sheet glass, motor vehicles, armaments, chemicals, ceramics, wood, paper products, footwear
Agriculture: accounts for 7% of GNP (includes forestry); largely self-sufficient in food production; diversified crop and livestock production, including grains, potatoes, sugar beets, hops, fruit, hogs, cattle, and poultry; exporter of forest products
Economic aid: donor—$4.2 billion in bilateral aid to non-Communist less developed countries (1954-89)
Currency: koruna (plural—koruny); 1 koruna (Kc) = 100 haleru
Exchange rates: koruny (Kcs) per US$1—27.65 (January 1991), 17.95 (1990), 15.05 (1989), 14.36 (1988), 13.69 (1987), 14.99 (1986), 17.14 (1985)
Fiscal year: calendar year

APPENDIX D - KEY CONTACTS: POLAND

This appendix was reprinted from "Poland: The New Business Frontier," March 1992. It was made available from the Eastern Europe Business Information Center, U.S. Department of Commerce, Room H-7412, Washington, DC 20230.

I. POLISH COMMERCIAL AND ECONOMIC OFFICERS IN THE UNITED STATES OF AMERICA

Embassy of the Republic of Poland
Office of the Commercial Counsellor
820 Second Ave. (17th floor)
New York, NY 10017
Telephone: (212) 370-5300-1-2, Telex: 595657 polcomer nyk
Fax: (212) 818-9623

Embassy of the Republic of Poland
Office of the Economic Counsellor
2640 16th Street, N.W.
Washington, DC 20009
Telephone: (202) 232-4517, Telex: 089480
Fax: (212) 328-2152

Consulate General of the Republic of Poland
Commercial Section
Ontario Street, Suite 39-06B
Chicago, IL 60611
Telephone: (312) 642-4102, Telex: 283042 polhan cgo

Consulate General of the Republic of Poland
3460 Wilshire Blvd., Suite 1200
Los Angeles, CA 90010
Telephone: (213) 395-4544
Fax: (213) 395-4544

II. COMMERCIAL AND ECONOMIC INSTITUTIONS, ORGANIZATIONS AND COMPANIES IN POLAND—GOVERNMENT INSTITUTIONS

Ministerstwo Wspolpracy Gospodarczej z Zagranics
Ministry of Foreign Economic Relations
Pl. Trzech Krzyzy 5, 00-507 Warszawa
Telephone: 693-50-00, Telex: 814501 mhz pl
Fax: 28-68-08

Ministerstwo Finansow
Ministry of Finance
ul. Swietokrzyska 12, 00-049 Warszawa
Telephone: 694-52-00, Telex: 815592, 814386 min. fin. pl

Minsterstwo Przemyslu
Ministry of Industry
ul. Wspolna 4, 00-505 Warszawa
Telephone: 21-03-51, Telex: 814267, 814261 mp pl

Ministerstwo Przekształcen Wlasnosciowych
Ministry of Ownership Changes
ul. Mysia 5, 00-496 Warszawa
Telephone: 28-32-61
Fax: 21-33-61

Ministerstwo Rolnictwa i Gospodarki Zywnosciowej
Ministry of Agriculture and Food Economy
ul. Wspolna 30, 00-159 Warszawa
Telephone: 21-03-11, Telex: 814597-98, 813685, 817188
Fax: 21-89-87, 29-68-96

Ministerstwo Lacznosci
Ministry of Telecommunication
Pl. Malachowskiego 2, 00-066 Warszawa
Telephone: 26-14-11, 26-10-71, Telex: 813001
Fax: 26-48-40

Ministerstwo Gospodarki Przestrzennej i Budownictwa
Ministry of Physical Planning and Construction
ul. Wspolna 2, 00-926 Warszawa
Telephone: 21-03-51, Telex: 814411, 814416
Fax: 25-58-87

Ministerstwo Ochrony Srodowiska, Zasobow Naturalnych i Lesnictwa
Ministry of Environmental Protection, Natural Resources and Forestry
ul. Wawelska 52/54, 00-922 Warszawa
Telephone: 25-00-01, 25-40-01, Telex: 812816, 817157
Fax: 25-33-35, 25-39-72

Agencja d/s Inwestycji Zagranicznych
Foreign Investment Agency
ul. Chopina 1, 00-559 Warszawa
Telephone: 29-35-53, Telex: 817225
Fax: 21-84-27

COMPANIES ENTITLED BY THE MINISTRY OF FINANCE TO MAKE AUDITING OF JOINT VENTURES AND FOREIGN PARTIES

Arnold Hill Sp. z o.o.
ul. Sikorskiego 35
04-476 Warszawa

Baltic Accountants and Consultants Ltd Sp. z o.o.
ul. Jankiela 8
Warszawa

Biuro Ekspertyz Finansowych i Consultingu PRO-INVEST Sp. z o.o.
Financial and Economic Consultants
ul. Mokotowskka 13 m. 1a
00-640 Warszawa
Telephone: 25-48-86, 25-55-03

"Book-Keeping-Services" Sp. z o.o.
Al. Jerozolimskie 65/79
Hotel Marriott
00-697 Warszawa

CAM Sp. z o.o.
ul. Kopernika 30
00-328 Warszawa

"DRT Poland" Sp. z o.o.
Al. Ujazdowskie 6 A
00-461 Warszawa
Telephone: 29-32-85, 29-32-81 ext. 215
Fax: 29-01-07

Ernst and Young TKD S.A.
ul. Wspolna 62
00-687 Warszawa

Eureka-Consulting, Poznanski Dom Emisyjny Sp. z o.o.
Poznan House of Issue
ul. Marcinkowskiego 22
61-827 Poznan
Telephone: 52-81-21, Telex: 413780

Ginter Janaszek Silesian Auditor Counsellor Sp. z o.o.
ul. Marchlewskiego 5
40-129 Katowice

International Business Consulting Sp. z o.o.
Al. Jerozolimskie 65/79
Hotel Marriott
00-697 Warszawa
Telephone: 29-72-32

KPMG-Reviconsult Sp. z o.o.
Al. Jerozolimskie 65/79
Hotel Marriott apt. 3603
00-697 Warszawa

Moore Stephens Sp. z o.o.
Al. Jerozolimskie 65/79
Centrum LIM
00-697 Warszawa
Telephone: 30-65-04, 30-65-05, Telex: 825366
Fax: 30-65-78

Price Waterhouse Poland Sp. z o.o.
ul. Bagatela 14
00-585 Warszawa

Przedsiebiorstwo Eksportowo-Importowe Produkcji, Handlu i Uslug "Abex" Sp. z o.o.
Export-Import Enterprise for Production, Trade and Services "Abex"
ul. Urszuli 18
65-147 Zielona Gora

Spoldzielcze Przedsiebiorstwo Promocji Postepu
Naukowo-Technicznego i Doradztwa Gospodarxcego "ORTEP"
Co-operative Enterprise for the Promotion of Scientific-Technological Progress and Economic Consulting
ul. Sienkiewicza 85/87
90-057 Lodz
Telephone: (0-42) 36-83-83, 36-56-00 ext. 446, 447
Telex: 886780

DATA BANKS ABOUT POLISH AND FOREIGN FIRMS LOOKING FOR PARTNERS FOR ECONOMIC CO-OPERATION

Agencja d/s Inwestycji Zagranicznych
Foreign Investment Agency
ul. Chopina 1
00-559 Warszawa
Telephone: 29-35-53, Telex: 817225
Fax: 21-84-27

Biuro Promocji Krajowej PIHZ-Centrum Promocji
Bureau for Domestic Promotion PIHZ-Promotion Centre
ul. Trebacka 4
00-074 Warszawa
Telephone: 26-73-76, 27-40-88, Telex: 814361
Fax: 27-46-73, 635-51-37

Biuro UNIDO w Warszawie
UNIDO Industrial Cooperation and Investment Promotion Service
ul. Stawki 2 (14th floor)
00-193 Warszawa
Telephone: 635-60-86, 635-71-12, Telex: 817916
Fax: 635-12-60

Business Foundation
Polska Fundacja Klubu Rzymskiego
Polish Foundation of the Club of Rome
ul. Wspolna 1/3
00-529 Warszawa
Telephone: 28-40-71 ext. 421, 456, 21-84-20, 21-99-93
Telex: 817008

Fundacja Gospodarcza NSZZ Solidarnosc
Foundation for Economy NSZZ Solidarnosc
Waly Piastowskie 24
80-855 Gdansk
Telephone: (0-58) 38-44-12, Telex: 514160, 513170
Fax: (0-58) 31-67-22, 31-07-17

Fundacja Polska
Polish Foundation
00-443 Warszawa
Telephone: 21-56-0, Telex: 817245
Fax: 28-16-66

C. Hartwig Gdansk
Przedsiebiorstwo Spedycji Miedzynarodowej
International Forwarding
ul. Na Zaspe 3
80-546 Gdanssk
Telephone: (0-58) 27-90-22, Telex: 054491
Fax:20-48-25

C. Hartwig Katowice
Przedsiebiorstwo Spedycji Miedzynarodowej
International Forwarding
ul. Slowackiego 37
40-093 Katowice
Telephone: (832) 53-92-41 to 49, Telex: 0315761
Fax: 832) 59-99-11

C. Hartwig Szczecin
Przedsiebiorstwo Spedycji Miedzynarodowej
International Forwarding
ul. Malczewskiego 5/7
71-616 Szczecin
Telephone: (0-91) 24-00-51, Telex: 0422111
Fax: (0-91) 89-666

C. Hartwin Warszawa
Przedsiebiorstwo Spedycji Miedzynarodowej
International Forwarding
ul. Poznanska 15
00-680 Warszawa
Telephone: 29-60-31 to 39, Telex: 814603, 813610
Fax: 29-14-01

CUSTOMS, INSPECTION AND FORWARDING COMPANIES

Glowny Urzad Cel
Central Customs Office
ul.Swietokrzyskaa 12
00-049 Warszawa
Telephone: 20-03-11, Telex: 814427

Urzad Patentowny Rzeczypospolitej Polskiej
Patent Office of the Republic of Poland
Al. Niepodleglosci 188
02-555 Warszawa
Telephone: 25-80-01, Telex: 813716 cint pl

Polcargo Consulting
Rzeczoznawstwo i Kontrola Towarow w Obrocie Miedzynarodowym
International superintendents and Testing Services
ul. Zeromskiego 32
81-369 Gdynia
Telephone: (0-58) 20-53-71, Telex: 54247
Fax: (0-58) 21-68-19

Shipcontrol
Przedsiebiorstwo Rzeczoznawstwa i Kontroli Ilosciowej
Exports and Supervisors Shipcontrol
ul. Polska 21
81-334 Gdynia
Telephone: (0-58) 20-70-96, Telex: 54271

Centralny Inspektorat Standaryzacji (CIS)
Quality Inspection Office
ul. Zurawia 32/34
00-515 Warszawa
Telephone: 21-64-21, Telex: 813653

Eastern Placement Inc.
Warsaw division
ul. Krzywickiego 34
02-078 Warszawa
Telephone: 28-92-61, Telex: 813517

Urzed Miasta Stolecnego Warszawy
Wydzial Promocji Gospodarczej
The Office of the Capital City of Warsaw
Economic Promotion Department
Pl. Bankowy 3/5
00-142 Warszawa
Telephone: 20-27-53, 20-34-83, 20-16-46, Telex: 817201
Fax: 20-11-78

BUSINESS INFORMATION AGENCIES

Infocredit
ul. Smolna 40, pol. 220
00-920 Warszawa
Telephone: 26-61-78, 26-81-84
Telex: 817841

Biuro Promocji Krajowej PIHZ—Centrum Promocji
Bureau for Domestic Promotion PIHZ—Promotion Centre
ul. Trebacka 4
00-074 Warszawa
Telephone: 26-73-76, 27-40-88, Telex: 814361
Fax: 27-46-73, 635-51-37

Fundacja Polska Eksportuje
Foundation for Polish Export
ul. Krolewska 27
00-060 Warszawa
Telephone: 27-68-10

Bank Informacji Gospodarczej TWIG
TWIG Data Bank
ul. Mazowiecka 13
00-052 Warszawa
Telephone: 27-53-15, 26-76-32, Telex: 814738
Fax: 27-15-51

Biuro Studiow i Doradztwa Gospodarczego "PROMASZ:
Bureau for Study and Economic Consulting
ul. Barbary 1
00-686 Warszawa
Telephone: 28-31-59, 29-26-16, Telex: 813815, 813923

Instytut Organizacji Przemyslu Maszynowego "ORGMASZ"
Buiro Organizacji Wspolpracy z. Kapitalem Zagranicznym
Institute for the Organization of Machine-Building Industry
Bureau for the Organization of Co-operation with Foreign Capital
ul. Zelazna 87
00-879 Warszawa
Telephone: 24-09-18, 24-60-61, Telex: 813747

Izba Przemyslowo-Handlowa Inwestorow Zagranisznych
Foreign Investors' Chamber of Industry and Commerce
Krakowskie Przedmiescie 47/51
00-071 Warszawa
Telephone: 39-19-03, 27-22-34
Fax: 26-85-93

DOING BUSINESS IN POLAND: USEFUL CONTACTS IN POLISH GOVERNMENT

Ministry of Privatization
ul. Krucza 36
00-049 Warsaw
telephone: 48/2/628 02 81
fax: 48//2/625 11 14
Tomasz Gruszecki, Minister
Contact: Ms. Joanna Debska
Department of Foreign Cooperation
telephone: 48/2/628 9531

Ministry of Finance
ul. Swietokrzyska 12
00-049 Warsaw
telephone: 48/2/694 52 00
fax: 48/22/26 63 52
Andrzej Olechowski, Minister
Contact: Mr. Krzysztof Link
Department of Foreign Cooperation
telephone: 48/22/26 65 44

Ministry of Industry and Trade
ul. Wspolna 4
00-505 Warsaw
telephone: 48/22/21 03 51
fax: 48/2/628 17 58
Andrzej Lipko, Minister
Contact: Mr. Piotr Kowalski
Department of Foreign Cooperation
telephone: 48/2/6 28 21 41
fax: 48/22/29 50 43

Ministry of Foreign Economic Relations
Pl. Trzech Krzyzy 5
00-507 Warsaw
telephone: 48/2/693 50 00
fax: 48/2/628 68 08
Adam Glapinski, Minister
Contact: Mr. Zbigniew Kubacki
U.S. Desk Officer
telephone: 48/2/6 93 58 58
fax: 48/22/29 06 17

Ministry of Agriculture and Food Economy
ul. Wspolna 30
00-519 Warsaw
telephone: 48-2-623 10 00, 623 20 00
fax: 48/22/21 89 87
Gabriel Janowski, Minister

Ministry of Telecommunications
Pl. Malachowskiego 2
00-066 Warsaw
telephone: 48/22/26 14 11, 26 10 71
fax: 48/22/26 48 40
Marek Rusin, Minister

Ministry of Housing and Construction
ul. Wspolna 2
00-926 Warsaw
telephone: 48/22/21 03 51
fax: 48/22/25 58 87
Andrzeg Diakonow, Minister

Minister of Environmental Protection
ul. Wawlska 52/54
00-922 Warsaw
telephone: 48/22/21 03 51
fax: 48/22/25 33 35
Stefan Kozlowski, Minister

Ministry of the Interior
ul. Rakowieska 2B
02-514 Warsaw
telephone: 48/22/21 02 51
Antoni Macierewic, Minister

For permit to purchase real estate:
Social Administrative Department
Telephone: 48/2/6 01 18 69

Ministry of Transportation
Department of Foreign Cooperation
ul. Chalubinskiego 4/6
00-928 Warsaw
telephone: 48/22/24 45 00
fax: 48/22/21 99 68
Ewaryst Waligorski, Minister

Ministry of Health and Social Policy
Department of Foreign Cooperation
ul. Mildowa 15
00-923 Warsaw
telephone: 48/22/31 74 08
fax: 48/22/31 21 44
Marian Miskiewicz, Minister

APPENDIX E - KEY CONTACTS: HUNGARY

This appendix was made available by the U.S. Department of Commerce, International Trade Administration, Washington, DC 20230.

This list contains contact information for U.S. Government, Hungarian government and multilateral agencies and organizations that can provide assistance to companies interested in trading with or investing in Hungary. For information about private companies offering business-related services in Hungary, contact the Commerce Department's Eastern Europe Business Information Center (EEBIC) at (202) 277-2645.

THE EASTERN EUROPE BUSINESS INFORMATION CENTER (EEBIC)

U.S. Department of Commerce
Room 7412
Washington, DC 20230
telephone: (202) 377-2645
fax: (202) 377-4473

[EEBIC is the central clearinghouse within the U.S. Government for information on doing business in Eastern Europe; publishes newsletter called "Eastern Europe Business Bulletin," as well as "Eastern Europe Looks for Partners"]

UNITED STATES FOREIGN COMMERCIAL SERVICE (FCS BUDAPEST)

Gary Gallagher, Senior Commercial Officer
Patrick Hughes, Commercial Officer
telephone: (36-1) 122-8600; telex: 22-7136
fax: (36-1) 142-2529

(for visits and appointments)
1062 Budapest, Hungary
Bajza utca 31 (near corner of Andrassy utca)

(for mail from the USA)
American Embassy—Budapest
U.S. Foreign Commercial Service
APO AE 09213-5270

(for international packages and "overnight" mail)
American Embassy (FCS)
1054 Budapest, Hungary
Szabadsag ter 12

[The U.S. Department of Commerce's Foreign Commercial Service officers stationed in the U.S. Embassy in Budapest can advice and assist American companies while they are in Hungary; FCS also operates American pavilions at Hungarian trade fairs.]

American Chamber of Commerce—Hungary
1068 Budapest, Hungary
Dozsa Gyorgy ut 84/a
Room 412
telephone: (36-1) 142-8752
fax: (36-1) 122-8890
Steven Bina, President (and Director, Schwinn Bicycle/Hungary)
Peter Fath, Executive Director

[Membership consists of over 125 U.S. companies doing business in Hungary; publishes newsletter called "AmCham Today."]

U.S.-Hungarian Business Council
c/o U.S. Chamber of Commerce
1615 H Street, N.W.
Washington, DC 20062
telephone: (202) 463-5473
fax: (202) 463-3114
Thaddeus Kopinski, Executive Director
Kay Larcom, Associate Director

[Activities include lobbying on commercial policy issues, promoting private enterprise in Hungary, and providing business facilitation services for U.S. and Hungarian member companies.]

U.S. Overseas Private Investment Corporation (OPIC)
1615 M Street, NW
Washington, DC 20527

Loans and Loan Guaranties
John C. Pazartzis, Associate Investment Officer
telephone: (202) 457-7198
fax: (202) 331-4234

Central and Eastern European Growth Fund
Graham Williams, Manager, Special Finance Projects
telephone: (202) 457-7105
fax: (202) 223-3824

Insurance
Elizabeth Keck, Insurance Officer
telephone: (202) 457-7066
fax: (202) 872-9306

Preinvestment Services (Opportunity Bank)
Eric Luhmann, Investor Services Officer
telephone: (202) 457-7010 or 1-800-424-6742
fax: (202) 331-4234

[OPIC is a U.S. Government organization that provides financing and political risk insurance for U.S. investments in developing countries, including Hungary. Also maintains a database of investment opportunities called the "Opportunity Bank" and organizes investment missions to eligible countries.]

International Finance Corporation (IFC)
Department of Investments, Europe
Room 19027
1850 Eye Street, N.W.
Washington, DC 20433
telephone: (202) 473-0575
fax: (202) 334-8965
Zijay K. Chaudhry, Division II Manager

[IFC is a World Bank affiliate that can provide financing for joint venture companies in Hungary; IFC will open an office in Budapest in early 1991.]

Multilateral Investment Guarantee Agency (MIGA)
1818 H Street, NW
Washington, DC 20433
telephone: (202) 473-5419
fax: (202) 679-0509
Edward A. Coppola, Regional Manager for Europe

[MIGA is a World Bank affiliate that can provide political risk insurance for joint venture companies in Hungary.]

U.S. Trade and Development Program (USTDP)
Washington, DC 20523
telephone: (703) 875-4357
fax: (703) 875-4009
Fred Eberhardt, Director for Europe

[USTDP is a U.S. government organization that funds feasibility studies for infrastructural projects in developing countries, including Hungary.]

Hungarian-American Enterprise Fund (3 offices)
1620 Eye Street, N.W.
Washington, DC 20006
telephone: (202) 467-5444
fax: (202) 467-5469
Alexander Tomlinson, President and CEO

1053 Budapest
Kecskemeti ut 10-12
Room 115
telephone: (36-1) 117-7598
fax: (36-1) 118-3821
Charles Huebner, Managing Director

[The Hungarian-American Enterprise Fund is a privately-managed fund that was established by the U.S. government through the 1989 "Support for East European Democracy" (SEED) Act. The fund's primary purpose is to promote the development of the Hungarian private sector by, among other things, investing in joint ventures between American and Hungarian companies.]

Embassy of the Republic of Hungary Commercial Office
2401 Calvert Street, N.W.
Suite 1021
Washington, DC 20008
telephone: (202) 387-3191; telex: 41-5902
fax: (202) 387-3140
Ferenc FURULYAS, Minister and Head of Hungarian Commercial Offices in the United States
Barnabas Nemeth, Commercial Counselor

150 East 58th Street, 33rd Floor
New York, NY 10022
telephone: (212) 752-3060; telex: 42-2438
fax: (212) 486-2958
Dr. Tibor NEMES, Commercial Counsellor

130 East Randolph Drive
Prudential Plaza, Suite 1930
Chicago, IL 60601
telephone: (312) 856-0274
fax: (312) 856-1080
Ivan NOVAK, Commercial Counsellor

[The Hungarian government's commercial offices in the United States can assist and provide information to American companies interested in investing in Hungary.]

National Bank of Hungary (NBH)
1850 Budapest, Hungary
Szabadsag ter 8-9
telephone: (36-1) 153-3044; telex: 22-7267
fax: (36-1) 153-1058
Dr. Bela SANDOR, General Manager, International Banking

10 Rockefeller Plaza, Suite 1100
New York, NY 10020
telephone: (212) 969-9270
Imre HOLLAI, Chief U.S. Representative

[The NBH can provide information on financial issues related to operating a joint venture company in Hungary.]

InvestCenter - TradeInform
(Hungarian Company for Investment and Trade Promotion)
1051 Budapest, Hungary
Dorottya utca 4
telephone: (36-1) 118-8220, 6064; telex: 22-5191
fax: (36-1) 118-3732
Ms. Gabriela Falus, General Manager

[Administered by the Hungarian Ministry of International Economic Relations; will set up appointments for foreign companies with appropriate Hungarian companies and government agencies, and can assist foreign companies in all aspects of establishing a joint venture in Hungary.]

Hungarian Chamber of Commerce and Industry
1055 Budapest, Hungary
Kossuth Lajos ter 6-8
telephone: (36-1) 153-3333, 153-3853
fax: (36-1) 153-1285
Tibor ZSELINSZKY, Officer for U.S. and Canada

[Represents Hungarian commercial interests; can assist foreign companies find and evaluate investment partners in Hungary.]

Ministry of International Economic Relations
(formerly the Ministry of Trade)
1055 Budapest, Hungary (street address)
1880 Budapest, Hungary (mailing address)
Honved U. 13-15
telephone: (36-1) 153-0000
fax: (36-1) 153-2794, 131-4186

Investment and Trade Promotion Agency (ITPA)
telephone: (36-1) 153-0000; telex: 22-5578
fax: (36-1) 153-2794
Mr. Zoltan GOMBOCZ, Director and Deputy State Secretary

[The ITPA was recently set up within the Ministry of International Economic Relations to assist foreigners in all aspects of trading with or investing in Hungary.]

Licensing Department
telephone: (36-1) 111-8433; telex: 20-2802
fax: (36-1) 153-2794
Mr. Janos CSENDES, Director of Licensing

[Issues import and export licenses for certain goods.]

Ministry of Finance
Main Department of International Financial Affairs
1051 Budapest, Hungary
Jozsef nador ter 2/4
telephone: 36-1-118-5670, 118-2066; telex: 22-5942
fax: 36-1-118-2570
Mr. Csaba REPASSY, Department Head

[Previously, this department of the Ministry of Finance issued licenses for joint venture companies that are more than 50% foreign-owned. As of amendments to the Foreign Investment Act that went into effect on January 1, 1991, the licensing requirement for all joint ventures has been abolished. However, this office may still serve as a useful investment-related contact.]

Ministry of Industry and Domestic Trade
(formerly the Ministry of Industry)
1024 Budapest, Hungary
85 Martirok utja
telephone: (36-1) 156-5566; telex: 22-5376
fax: (36-1) 175-3636, 155-3482
Mr. Zoltan LEHEL, Director, International Affairs

Ministry of Transport, Communication and Water Management
(formerly the Ministry of Transport, Communications and Construction)
1400 Budapest, Hungary
P.O. Box 87
Dob utca 75-81
telephone: (36-1) 122-0220, 122-9478; telex: 22-5729
fax: (36-1) 122-8695
Mr. Bela DOROS, Deputy State Secretary for Telecommunications
Mr. Gyula PARTOS, Director, Telecommunications Policy
Ms. Eva MOLNAR, Director, International Affairs

Ministry of Agriculture and Food
1055 Budapest
Kossuth Lajos ter 11
telephone: (36-1) 153-3000

Hungarian Telecommunications Company (HTC)
(formerly the "PTT")
1541 Budapest, Hungary
Krisztina krt 6-8
telephone: (36-1) 175-0121; telex: 22-1193
fax: (36-1) 156-0855
Pal HORVATH, Director General Andras GYORI, Manager, Business Relations

[The HTC operates Hungary's national telecommunications system.]

State Property Agency (SPA)
1051 Budapest, Hungary
Vigado utca 6
(Letter address: 1399 Budapest, Pf. 708)
telephone: (36-1) 118-5044
fax: (36-1) 118-7155
Lajos CSEPI, Director Karoly SZABO, Deputy Director

[Created in late 1989 to administer the Hungarian government's program of privatizing state-owned enterprises. The SPA could play a role in the government's decision-making process in cases where a foreign investment involves or results in the privatization of a state-owned firm.]

State Property Agency Information Office
1074 Budapest, Hungary
Harsfa utca 21
telephone: (36-1) 142-7734; telex: 22-6564
fax: (36-1) 142-7337

[The SPA Information Office disseminates information to the public regarding privatization in Hungary.]

HungExpo
1441 Budapest, Hungary
P.O. Box 44
Dobi Istvan ut 10
telephone: (36-1) 157-3555; telex: 22-4684 hexpo h
fax: (36-1) 128-5034

[Manages international commercial fairs and exhibitions at the Budapest fairgrounds. The U.S. Department of Commerce operates American pavilions at several of these fairs.]

Association of Industrial Cooperatives (OKISZ)
1146 Budapest, Hungary
Thokoly ut 58-60
telephone: (36-1) 141-4752; telex: 22-7576
fax: (36-1) 141-5521
Lajos KOVESKUTI, President

[Represents the interests of the Hungarian industrial cooperatives]

Association of Entrepreneurs (VOSZ)
1118 Budapest, Hungary
Szuret ut 15
telephone: (36-1) 185-3377; telex: 22-3436
fax: (36-1) 182-0316
Janos PALOTAS, President

[Represents the interests of the Hungarian small business sector.]

Foundation for Small Enterprise Economic Development (SEED)
1119 Budapest, Hungary
Andor utca 60
tel: (36-1) 181-0590
fax: (36-1) 181-2959
Dr. Agnes TIBOR, Managing Director

[Promotes the development of small business in Hungary.]

Central Corporation of Banking Companies
(Penzintezeti Kozpont, or "PK")
Foreign Exchange and Supervisory Office
(Deviza Engedelyezesi es Ellenorzesi Osztaly)
1093 Budapest
Szamuely u 38
telephone: (36-1) 117-1255
fax: (36-1) 118-8149

[Government office that can assist foreign companies in opening representational offices, securing office space and telephone service, hiring staff, as well as other practical business matters. PK is also involved in approvals for real estate acquisitions by foreign firms in certain cases.]

State Taxation Office (APEH)
1096 Budapest, Hungary
Haman Kato ut 3-5

[APEH administers Hungary's taxation system; joint venture companies must register the sectorial department of APEH that corresponds to the company's line of business; APEH's sectorial departments and their telephone numbers are listed on p. 18 of this report.]

Budapest Court of Firms
(i.e. the "Court of Registration")
1064 Budapest, Hungary
Rozsa Ferenc utca 79
telephone: (36-1) 112-1629

[Joint venture companies must register with the Court of Firms in the municipality or county of operation.]

Other Courts of Firms are located in these county seats and municipalities:

County (Megyei)	Capital (county seat)
Bacs-Kiskun	Kecskemet
Baranya	Pecs
Bekes	Bekescsaba
Borsod-Abauj-Zemplen	Miskolc
Csongrad	Szeged
Fejer	Szekesfehervar
Gyor-Moson-Sopron	Gyor
Hajdu-Bihar	Debrecen
Heves	Eger
Jasz-Nagykun-Szolnok	Szolnok
Komarom-Esztergom	Tatabanya
Nograd	Salgotarjan
Pest	Budapest
Somogy	Kaposvar
Szabolcs-Szatmar-Bereg	Nyiregyhaza
Tolna	Szekszard
Vas	Szombathely
Veszprem	Veszprem
Zala	Zalaegerszeg

Municipality (Megyei Varos)	Population (1988)
Budapest	2,104,700
Debrecen	217,364
Miskolc	209,807
Szeged	187,800
Pecs	181,356
Gyor	130,703

Office of Foreigners' Affairs (REOKH)
Andrassy ut 12 (formerly Nepkoztarsasag ut 12)
1061 Budapest, Hungary
telephone: (36-1) 118-0800

[Foreigners working in Hungary must obtain a residence permit from KEOKH.]

National Office of Inventions (Orszagos Talalmanyi Hivatal)
Patent Information Center (Szabadalmi Informacios Kozpont)
1370 Budapest, Hungary
P.O. Box 552
telephone: (36-1) 112-4400
fax: (36-1) 131-2596

[Provides information and assistance on patent-related issues.]

Hungarian Chamber of Commerce Arbitration Court
1055 Budapest, Hungary
Kossuth Lajos ter 6-8
telephone: (36-1) 153-2425

APPENDIX F - KEY CONTACTS: CZECHOSLOVAKIA

This appendix was made available by the U.S. Department of Commerce, International Trade Administration, Washington, DC 20230. Dated March 1992.

There are three relevant governments in Czechoslovakia: the Federal Government, in Prague; the government of the Czech Republic, in Prague; and the government of the Slovak Republic in Bratislava. Under a new so-called competency law the two constituent republics will be responsible for their economic development, and the Federal Government will create conditions for a single market. It extends significant economic powers to the republics, particularly in the area of privatization. For this reason, and because of the political nature of the competency law, companies visiting Czechoslovakia for the first time are strongly advised to visit both Prague and Bratislava to view business opportunities firsthand. For additional assistance, contact the U.S. Embassy Prague, Commercial Section, Robert Shipley or Shelley Galbraith; telephone (42/2) 536641 or 532470; telefax (42/2) 532457 or 537534.

To arrange appointments with ministries, companies are advised that the volume of business visitors to Prague each week does not enable the U.S. Embassy to request appointments on behalf of business representatives in all cases. Companies should first seek advice on appropriate government contacts, attempt to schedule appointments directly, and then seek further assistance from the U.S. Embassy, as warranted. Companies are also advised that because of ongoing changes in the governments in Czechoslovakia, this information may not be current. This information represents the best available as of the date listed above.

FEDERAL GOVERNMENT

President Vaclav Havel
Office of the President
10000 Prague 1 (Hrad)
Telephone: (42/2) 2101; Telex: 121911
Fax: (42/2) 535 674
Contact: Chief of Staff Mr. Karl Schwartzenberg
Mr. Richard Wagner, Economic Advisor to the President
Telephone: (42/2) 536 058

Minister Vladimir Dlouhy
Federal Ministry of Economy
Nabrezi kpt. Jarose 1000
17 0 3 2 Prague 7
Telephone: (42/2) 389 1111; Telex: 121044
Fax: (42/2) 375659 (Dlouhy only)

Prime Minister Marian Calfa
Presidium of the Government
Nabrezi kpt. Jarose 4
11000 Prague 1
Telephone: (4 2/ 2) 210 2; Telex: 121476
Fax: (42/2) 231 1446
Mr. Vaclav Vales
Deputy Prime Minister—Economy
Contact: Mr. Emilian Vos
Mr. Jiri Stadler

Minister Vaclav Klaus
Federal Ministry of Finance
Letenska 15
11810 Prague 1
(42/2) 514111
Telex: 121868
Fax: (42/2) 534498

Minister Jozef Baksay
Federal Ministry of Foreign Trade
Politickych veznu 20
11249 Prague 1
Telephone: (42/2) 2126; Telex: 121299
Fax: (42/2) 232 2868, 232 2861
Contact: USA Desk
Telephone: (42/2) 2126 2112, ext. 2132

Minister Josef Vavrousek
Federal Ministry of the Environment
Slezska 9
12029 Prague 2
Telephone: (42/2) 256488; Telex: 121427 or 121107
Fax: (42/2) 215 2659
Contact: Deputy Minister Mikolas or Mr. Dobes, Foreign Relations
Telephone: (42/4) 215 2422

Minister Lubos Dobrovsky
Ministry of Defense

Minister Emil Erenberger
Federal Ministry of Posts and Telecommunications
Klimentska 27
12502 Prague 1
Telephone: (42/2) 235 5545, 232 3924; Telex: 111410
Fax: (42/2) 236-8379
Contact: Ivan Laska or Ales Michna, Deputy Ministers

Minister Pavel Hoffman
Federal Ministry for Strategic Planning
Nabrezi kpt. Jarose 1000
17032 Prague 1
Telephone: (42/2) 389 1111; Telex: 121044

Minister Jiri Nezval
Federal Ministry of Transport
Nabrezi L. Svobody 12
12511 Prague 1
Telephone: (42/2) 2122-1111; Telex: 121096
Fax: (42/2) 236 8379
Contact: Mr. Dusan Valasek
Director, Foreign Relations Dept.

Government of the Czech Republic

Petr Pithart, Premier
Czech Republic
(Urad vlady CR)
Lazarska 7
11000 Prague 1
Telephone: (42/2) 235 8812; Telex: 122724

Minister Bohumil Kubat
Ministry of Agriculture and Food
Tesnov 65
11006 Prague 1
Telephone: (42/2) 286 2111; Telex: 121 041
Contact: Ing. Josef Hruby
Chief, Foreign Relations
Telephone: (42/2) 286 2257

Minister Ivan Dejmal
Ministry of Environment
Trida SNB 65
10100 Prague 10
Telephone: (42/2) 712 1111; Telex: 122 298
Fax: (42/2) 731 357
Contact: Ing. Vladimir Prchlik or Ing. Libuse Deylova

Minister Vlast Stepova
Ministry of Trade and Tourism
Staromestske nam. 6
11000 Prague 1
Telephone: (42/2) 2897; Telex: 121513
Fax: (42//2) 232 6815
Contact: Dr. Karel Buchnar
Chief, Foreign Relations
Telephone: (42/2) 2897 ext. 421

Minister Karel Spacek
Ministry of Finance, Prices and Wages
Letenska 15
11000 Prague 1
Telephone: (42/2) 514 1111; Telex: 121868
Contact: Ing. Hynek Rasocha
Telephone: (42/2) 530 859

Minister Karel Dyba
Ministry of Economic Policy and Development
Trida SNB 65
10 010 Prague 10
Telephone: (42/2) 747198
Fax: (42/2) 739114
Contact: Zdenek Krulis, Director Foreign Relations
Telephone: (42/2) 712 1111

Minister Jan Vrba
Ministry of Industry
Na Porici 24
11180 Prague 1
Telephone: (42/2) 232 3777; 232 1282
Contact: Ladislav Kubicekk
Telephone: (42/2) 235 6499

Minister Martin Bojar
Ministry of Health
W. Piecke 98
10100 Prague 10
Telephone: (42//2) 2118; Telex: 121602 or 121048
Contact: Dr. Jiri Macke
Chief, Foreign Relations
Telephone: (42/4) 294766

Minister Tomas Jezek
Ministry for Management of State Property and Privatization
Letenska 15
11000 Prague 1
Telephone: (42/2) 514 2109; Telex: 121 868

Government of the Slovak Republic

Jan Carnogursky, Premier
Slovak Republic
(Urad vlady SR)
Nam. svobody 1
81370 Bratislava
Telephone: (42/7) 45584
Fax: (42/7) 315 484
Anton Vavro, Deputy Premier for Economic Reform

Minister Ivan Tirpak
Ministry of the Interior and Environment
Hlboka c. 2
81235 Bratislava
Telephone: (42/7) 42451
Fax: (42/7) 311 384
Contact: Ing. Vladimir Jurik

Minister Josef Krsek
Ministry of Agriculture and Food Suvorova 12
80000 Bratislava
Telephone: (42/7) 59721; Telex: 93354
Fax: (42/7) 57834

Minister Jozef Chren
Ministry of Trade and Tourism
Cs. Armady 8
81315 Bratislava
Telephone: (42/7) 300; Telex: 93192
Contact: Drahoslav Machac

Minister Viliam Oberhauser
Ministry of Forests and Water Economy
L. Novomeskeho 2
80000 Bratislava
Telephone: (42/7) 325303; Telex: 93381
Fax: (42/7) 328968
Contact: Ing. Viktor Drozda
Telephone: (42/7) 326308

Minister Jozef Belcak
Ministry of Economy
Drienova 24
82611 Bratislava
Telephone: (42/7) 234332; Telex: 92349
Fax: (42//7) 233595
Contact: Alzbeta Durankova
Telephone: (42/7) 46395
Fax: (42/7) 43476

Minister Jan Holcik
Ministry of Industry
Mileticova 1
82400 Bratislava
Telephone: (42/7) 374427; Telex: (42/7) 92543
Fax: (42/7) 376 364
Contact: Ing. Dusan Gbelsky
Telephone: (42/7) 374427

Minister Jozef Butora
Ministry of Construction
Stefanovicova 3
81325 Bratislava
Telephone: (42/7) 47541; Telex: 92548
Fax: (42/7) 44727

Minister Alojz Rakus
Ministry of Health and Social Matters
Cs. armady 6
81305 Bratislava
Telephone: (42/7) 50946; Telex: 92361
Contact: Milos Ihnat

Minister Jozef Danco
Ministry of Finance, Prices and Wages
Stefanovicova 3
80000 Bratislava
Telephone: (42/7) 46952
Fax: (42/7) 48042

Minister Ivan Miklos
Ministry for Privatization and Administration of National Property
Kycerskeho 1
80000 Bratislava
Telephone: (42/7) 46951
Fax: (42/7) 43476

Contacts: Finding An Agent Or Distributor

State agencies representing foreign firms:

USOZ/Ucelove sdruzeni organizaci pro zahranicni zastoupeni
(Association of Czechoslovak Foreign Trade Agencies)
Na strizi 63
14062 Prague 4
Telephone: (42/2) 414 1111, 437 040, 438 632; Telex: 121 355
Fax: (42/2) 434 690
Contact: Ludmila Kikarova

Efecktim
Vaclavske namesti 1
11334 Prague 1
Telephone: (42/2) 236 8359; Telex: 121 117
Fax: (42/2) 232 0094

Represents foreign firms in capital goods for engineering and chemicals industries; for the textile, leather, paper, foods, ceramics and business industries; for data processing, electronics, medical equipment, mining, metallurgical, transport and drilling technology.

Interal
Pionierska 9/A
81345 Bratislava III
Telephone: (42/7) 65103, 65387; Telex: 93263
Fax: (42/7) 824312

Represents foreign firms in machine tools and equipment for the production of electric motors, mining technology, woodworking and papermaking machinery, food industry machinery, pesticides, plastics, water treatment technology, packaging machines, technology for construction engineering, bearings, hydraulics and electronics.

Intersim
Topolova 2916/14
10000 Prague 10
Telephone: (42/2) 758331, 757742; Telex: 121963, 121971
Fax: (42/2) 757510

Represents foreign firms in machine tools, machinery and equipment for the woodworking, food, ceramic and textile industries; equipment for metallurgical plants, foundries ad power engineering; engineering instruments, material testing, measuring and controlling equipment; consumer electronics, computers, passenger cars and road building and general building machinery.

Media
Strakonicka 510
15063 Prague 5
Telephone: (42/2) 545346; Telex: 121925
Fax: (42/2) 546 741

Represents foreign firms in the areas of heavy and light engineering; air conditioning; medical technologies; lasers; textile machinery; measuring and control instruments; woodworking machines; computers; synthetic flooring; insulating and sealing pastes and plasters; consumer goods; food, drink and raw materials for food and drink; production of microelectronics; agricultural equipment; packaging materials; electro-optics.

Mercanta
Balbinova 22
12034 Prague 2
Telephone: (42/2) 236 0921, 246008; Telex: 122789, 121 728
Fax: (42/2) 232 2868, 232 2861

Represents foreign firms in the areas of engineering, transportation equipment; laundry and dry-cleaning machinery; water treatment equipment; respirators and fire-fighting and rescue equipment; industrial safety equipment, medical equipment; consumer goods; photographic and optical equipment.

Phoenix
Bratri Capku 32
10000 Prague 10
Telephone: (42/2) 426 213, 426 212; Telex: 62682, 52504
Fax: (42/2) 692 2902

Represents foreign firms in equipment and machines for construction and earth moving, welding, transportation, packaging and metallurgy; textile machines.

Pragent
Wasingtonova 17
11000 Prague 1
Telephone: (42/2) 242151, 250703; Telex: 122713
Fax: (42/2) 231 5327

Represents foreign firms in consumer goods; restaurant and catering equipment; electric and manual tools; raw materials for the building, furniture, textile, leather, shoe, electrical products and automotive industries.

Rephachem
Kodanska 46
10100 Prague 10
Telephone: (42/2) 732 176; Telex: 121 905
Fax: (42/4) 732 035

Represents foreign firms in the chemicals, pharmaceuticals and related machinery industries.

Tradex
Jungmannova namesti 7
11000 Prague 1
Telephone: (42/2) 235 5693, 235 5694; Telex: 122 707
Fax: (42//2) 235 5698

Represents foreign firms in the fields of equipment for the engineering and metallurgy industries; conveyor belts, fork-lift trucks and materials handling equipment; hospital and laboratory instruments, medical equipment; cleaning technologies; consumer goods.

Unifrux
V jame 3/699
11345 Prague 1
Telephone: (42/2) 242213, 242 280, 235 0275; Telex: 121 944, 122996
Fax: (42/2) 231 6023

Represents foreign firms and provides intermediary services in the fields of foodstuffs and agricultural products; specialized chemical products, cosmetics, pharmaceuticals; machinery and equipment for the food industry; packing and packaging equipment; machines and tools for the furniture industry; tools for machine tools and metal finishing; microelectronics, parts, small computer technology and measuring instruments; power engineering and metallurgy equipment.

Zenit
Holubova 11
15000 Prague 5
Telephone: (42/4) 530311; Telex: 121801
Fax: 536293

Represents foreign firms in machinery and equipment for the engineering, power engineering, textile, building, pharmaceuticals, chemicals and foods industry; mechanized equipment for packaging and electronic control systems; controlling and measuring devices for the gas industry, oil refineries and other industries; chemicals, pharmaceuticals, dyes, additives for building materials, plastics and consumer goods.

APPENDIX G - ENVIRONMENTAL SUMMARY: POLAND

This appendix was made available by Eastern Europe Business Information Center, Eastern Europe Business Development Program, U.S. Department of Commerce, Room 7412, Washington, DC 20230.

I. THE ENVIRONMENTAL SECTOR: A BUSINESS OPPORTUNITY

Environmental pollution clean-up and control is considered Poland's top priority. Growing public awareness of the magnitude of environmental problems in Poland has placed further demands on the Government to implement strong environmental controls. In response, the Government of Poland has designated environmental pollution control and clean-up as a priority area for foreign investment and exports.

In conjunction with this focus on its improvement of environmental conditions, Poland has made significant strides to orient its centrally-planned economy to free market principles. The country's recent move to privatize government-owned enterprises now places buying decisions in the hands of individual enterprises. Although the foreign trade sector is in the process of being decentralized, U.S. companies should continue to establish contacts with the appropriate ministries. See Section IV for environmental contacts in Poland.

Poland's movement toward privatizing its economy and cleaning and protecting its environment offers U.S. companies unprecedented business opportunities. U.S. expertise in environmental technology and in energy management will contribute to solving Poland's problems in several ways.

- Pollution control equipment and expertise to help monitor and minimize the environmental impact of the continued reliance on brown coal;
- Clean-coal technology and pollution control equipment to retrofit existing lignite-fired coal plants in priority pollution-distressed areas of Southern Poland; and
- Expertise in water and soil clean-up and in waste management to improve public health conditions.

U.S. businesses interested in investment opportunities in these three areas are eligible for lower taxes and low-cost financing. Firms that invest in these preferred areas are exempt from "deposit" requirements. Exemptions from the State turnover sales tax are granted to those firms that use waste or recycled materials. Under the new law on joint ventures of December 24, 1989, special treatment is granted to U.S. firms in obtaining operating permits and tax holidays.

This overview of the environmental sector in Poland serves as a guide to U.S. companies looking to explore opportunities in this important sector. Section II provides background information on the environmental problems faced by Poland; Section III is a detailed outline of the government priorities in air, water, and soil pollution; Section IV is a comprehensive overview of U.S. government and other sources of financing; Sections V and VI list contacts in the environment in Poland and the U.S.; and Section VII provides a listing of trade promotional events in which U.S. companies may participate.

II. OVERVIEW OF THE ENVIRONMENT & ENERGY RESOURCES: POLAND

Four decades of central planning in Poland have left a legacy of severe contamination of air, water, and soil, with concomitant deterioration in public health. As a result, life expectancy is lower than twenty years ago. 10-15 percent of youth is chronically ill. Reported incidents of severe asthma, bronchitis, and allergies, even in young children, have been on the rise. The need for emergency corrective action is paramount for Poland's adjustment to a free-market economy and to political democracy.

The former system of central planning emphasized heavy, energy intensive industries, and was supported by extensive subsidies to the Polish coal industry. Coal is the major indigenous energy resource, accounting for 95 percent of primary energy production, and about 77 percent of energy consumption. Little or no efforts were made to conserve energy, either in industry or among the general population, because of the abundance and low price of hard coal. Of the roughly 250 million metric tons (mmt) of coal produced in 1989, about 178 mmt was bituminous hard coal, and the rest was lignite, or brown coal.

The bulk of the bituminous coal mined is relatively low sulfur (less than 1 percent sulfur). However, a complete lack of combustion controls and grossly inefficient operating practices by industry and residential users alike virtually ensures that major coal producing and consuming regions such as Upper Silesia (where 66 of 70 of the country's hard coal mines are located) are continually bathed in a stew of airborne particulates and by-products of incomplete combustion. The World Commission on Environment and Development has reported that in Katowice (a major mining and industrial center in the heart of Upper Silesia), safety limits are often exceeded for levels of sulfur dioxide by 200 percent; lead by 196 times; airborne dust by 35 times; nitrogen oxide by six times; and carbon monoxide by 21 times.

As a percent of energy-use per dollar of gross domestic product (GDP), Poland uses close to three times as much energy as Belgium whose level of GDP is higher. As a region, East European industry contributes only one third to Europe's industrial output but accounts for two thirds of its sulfur dioxide emissions. The lion's share of these sulfur dioxide emissions are from East Germany and Czechoslovakia where the lignite is unusually high in sulfur and there are no pollution controls. About half of the approximately 820 mmt of brown coal burned in Eastern Europe in 1988 (exclusive of the USSR) came from these two countries.

For the foreseeable future, Poland will rely on coal as its main source of energy. Its access to cleaner energy sources, oil and natural gas, is hampered by high world prices due to the Persian Gulf crisis and by the lack of diversification of its suppliers. The Soviet Union formerly supplied Poland with all of its oil through artificially-priced barter arrangements. Over the past year, the Soviet Union has drastically reduced oil shipments, owing to production disruption at home. Moreover, as of January 1, 1991 the Soviet Union will be asking for payments at world market prices and in hard currency.

Polish industries continue to dump industrial wastes and municipalities continue to discharge raw, untreated sewage into rivers and the Baltic Sea. As a result, the coastline along the Baltic Sea and a third of Polish rivers suffer from serious pollution. The levels of salinization of the Vistula and Oder rivers due to waste water discharge from mines are increasing. According to Polish authorities, 60 percent of the Vistula River is not fit for industrial use because it would clog and corrode industrial machinery.

Partly due to environmental fallout from air and water pollution and to accidental and intentional dumping of hazardous wastes, soil contamination and forest damage are reaching dangerous proportions. In the past, Poland has accepted shipments of toxic and hazardous waste in return for hard currency, but did not have facilities for safe disposal nor regulated storage. High concentrations of lead and other heavy metals makes it dangerous to grow crops on one-fourth of Poland's farmland.

Although new environmental legislation has been prepared, the regulations needed to implement these laws are currently being rewritten. Environmental officials anticipate that the forthcoming legislation will conform, to a large extent, with environmental standards of the European Community. U.S. environmental companies would do well to hold to these same standards when exporting to or investing in Poland.

The following section provides lists of designated priority areas of air, water, and soil clean-up and protection and corresponding technologies.

III. PRIORITY AREAS

The Government of Poland has designated the following areas in air, water, and soil clean-up and protection as priority areas for foreign investment and exports.

1) Air:

- Sulfur dioxide and particulate abatement;
- Desulfurization of lignite coal;
- Fluidized bed boilers;
- Retrofitting existing power plants to desulfurize emission gases, e.g., Silesia, Jaworzno, Rybnik, Laziska, Pilaniec, Skawina, Turow, and Belchatow;
- Thermal power plants and district-heating systems in Krakow, Katowice, Torun, Poznan, Wroclaw, and Lodz;
- Reduction of nitrogen oxide emissions;
- Introduction of unleaded petrol and catalytic converters;
- Coal preparation technology and process control for bituminous coal;
- Operator training programs for small- and industrialized coal-fired boilers for complete fuel combustion and reduction of nitrous oxide emissions;
- Process controls for complete combustion in small- and medium-sized boiler houses;
- Market and distribution training for coal producers;
- Coal emissions technologies geared for particulate control, baghouses, precipitators, and tri-ox burners; and
- Residential and industrial energy management technologies and controls for energy conservation.

Technologies needed: Technical assistance and equipment for investment projects, e.g., clean-coal technologies, desulphurization facilities for coal-fired power plants, automatic meters to measure pollutant emissions; efficient highly-toxic waste incineration plants with recovery of heat; monitoring systems for pollutants; process controls for preparation plants, boilers operations for complete combustion and reduction of nitrous oxides; particulate emissions control technologies and process controls for coal-fired plants of all sizes (precipitators, baghouses and tri-ox burners); computer software for financial investment analysis, accounting, technical graphics and mining engineering analysis (electrical and ventilation network analysis); and complete energy management/conservation technologies and controls for residential and industrial use.

2) Water:

- Potable drinking water for households in urban areas;
- Conservation and clean-up of Great Mazurian Lake system and the Baltic Sea coastal area;
- Installation of municipal wastewater treatment plants to remove biogenic compounds and to process sludge; and

- Desaltinization of wastewater from coal mines, especially, Czeczott, Piast, Ziemowit, Budryk, and Debiensko.

Technologies needed: Machinery and control equipment for wastewater treatment plants; design of management systems for water resources; mine water desalination plants, and equipment for treatment of raw sewage affecting non-artesian groundwater and wastewater: Aeration equipment using compressed air, press and centrifuge devices for dewatering of sludge, diving pumps, and automatic control for treatment of wastewater.

3) Soil/Forest:

- Production of equipment to collect, transport, process and store waste;
- Introduction of organic farming to provide supply of safer, high quality food for domestic and foreign markets;
- Recovery of metals from metal-borne waste/steelworks in Katowice and Krakow, and heavy metals metallurgy in Legnica, Olkusz, and Miasteczko Slaskie;
- Segregation of municipal waste with recovery of raw materials;
- Waste land recultivation and afforestration; and
- Protection of Polish National Parks and valuable recreational regions.

Technologies Needed: Technologies for the removal and storage of hazardous waste; operation of waste disposal collection facilities and landfills; programme for waste management and utilization, and recycling; construction of pilot waste incineration plants; establishment of protection zones for large industrial plants; recovery of raw materials from solid waste; and establishment of national parks.

IV. FINANCING

Commercial sources of financing may be limited since money center banks have been reluctant to lend overseas. U.S. companies, therefore, should be innovative in arranging financing. U.S. companies are encouraged to pursue the following financing opportunities, available through U.S.- and foreign-sponsored programs as well as contract possibilities at the various U.S. government, bilateral and multilateral agencies.

U.S. Government Sources of Financing:

* Overseas Private Investment Corporation (OPIC). OPIC promotes economic development by encouraging U.S. private investment in developing countries, including Eastern Europe. OPIC promotes investment through its financing and insurance programs. Lending is provided to joint venture projects, wherein the majority owner is a U.S. company.

Through its direct loans and loan guarantees, OPIC provides medium-to long-term financing for foreign investment. OPIC provides medium- to long-term financing for foreign investment. OPIC also insures investment projects against the risks of war, revolution, insurrection, civil strife, and inconvertibility of local currency.

For information on finance programs in Poland, contact:
Susan Goeransson
Overseas Private Investment Corporation
1615 M Street, NW
Washington, DC 20527
(202) 457-7183

For information on insurance programs in Poland, contact:
Christiane Carney
Overseas Private Investment Corporation
1615 M Street, NW
Washington, DC 20527
(202) 457-7050

The Opportunity Bank is a computer databank that matches U.S. investor interests with specific opportunities in Poland. This service is available for a nominal fee while registration is free. For further information, contact Eric Luhmann at (202) 457-7128.

* Export-Import Bank of the United States (EXIM). EXIM helps U.S. companies finance their exports abroad through a variety of loan, guarantee, and insurance programs. EXIM loans, backed by the full faith and credit of the U.S. Government, extends direct credit to foreign buyers of U.S. exports, which helps the U.S. exporter get paid in hard currency.

For further information on all EXIM's programs, contact:
Mr. William Browning
Export-Import Bank of the United States
811 Vermont Avenue, NW
Washington, DC 20571
Telephone: (202) 566-8862

For the West Coast office of EXIMBANK, please contact:
Mr. Art Obester
Export-Import Bank of the United States
811 Vermont Avenue, NW
Washington, DC 20571
Telephone: (213) 575-7425

* U.S. Trade and Development Program/Poland. The U.S. Trade and Development Program promotes economic development and U.S. exports by funding feasibility studies and other project planning services for major infrastructure and industrial projects. Grant assistance is provided for studies in high priority sectors, like solid and hazardous waste management, energy development, pollution control, and telecommunications. This assistance is provided to help Poland identify its investments needs. Inquiries may be addressed to the contact person below.

Mr. Fred Eberhart
Regional Director for Eastern Europe
U.S. Trade and Development Program
SA-16, Room 309
Washington, DC 20523-1602
Telephone: (703) 875-4357
Fax: (703) 875-4009

* U.S. Agency for International Development: Three major environmental projects on Eastern Europe are underway for fiscal year 1991.

1) **Environmental Initiatives Project** is designed to improve the indigenous capacity of countries to manage current and future environmental problems through strategic country

assessments, legal reform, and private sector initiatives on environmental health and safety.

2) **Environment Training Project** will provide environmental business management and administration training, public outreach, and scientific linkages to improve the human resource base.

3) **Improved Public Sector Environmental Services** is designed to improve the environmental quality and strengthen government capability through natural resources management, municipal government assistance, and clean-up efforts.

For further information on these projects and other U.S. AID projects, contact:

Ms. Lori Freer
Environment and Natural Resource Division
U.S. Agency for International Development
2201 C Street, NW
ENE/TR/ARD, Room 4440
Washington, DC 20523-0053
Tel: (202) 647-7805
Fax: (202) 647-6962

Other Sources of Financing:

* Polish-American Enterprise Fund. This Fund was established under the Support for East European Democracy Act as a private, non-profit corporation. This Fund is aimed to assist small- and medium-sized enterprises and U.S.-Polish joint ventures promote private sector development through grants, equity investments, insurance, guarantees, and training assistance. To discuss project guidelines and criteria, contact the Fund directly at the following telephone number and address below.

Mr. Robert G. Faris
President
Polish-American Enterprise Fund
535 Madison Avenue, 36th Floor
New York, NY 10022
Telephone: (212) 906-8660
Fax: (212) 758-9377

* Environmental Enterprises Assistance Fund. This Enterprises Fund, established by Winrock International and the U.S. Agency for International Development, will be capitalized at $20 million. The goal of the Fund is to help promote small-scale, commercially viable renewable energy and environmentally sound private enterprises in developing countries, like in Eastern Europe. Through technical and management assistance, and loans and equity investments, the Fund will assist small-scale technologies overcome barriers to market-entry.

Franklin Tugwell
Environmental Enterprises Assistance Fund
1611 North Kent Street
Suite 600
Arlington, VA 22209
Telephone: (703) 525-9430
Fax: (703) 243-1175

* The Foundation for Energy Efficiency (Energy-Efficiency Centers, Warsaw and Katowice). These Energy Efficiency Centers were established to promote the energy efficiency policy and technology development through 1) policy research and development; 2) the development of private joint ventures in efficiency technology and services; 3) training and demonstration projects; and 4) public education and data and information services. U.S. companies may look to these centers to find joint venture partners and locate international sources of funding for their projects.

In the United States, contact:
William U. Chandler
Battelle, Pacific Northwest Laboratory
901 D Street, SW
Suite 900
Washington, DC 20024-2115
Telephone: (202) 646-5242
Fax: (202) 646-5233

For the Warsaw Center, contact:
Adam Gula
Director of Policy and Outreach
ul. Gorskeigo 7-IIIp
00-033 Warszawa
Poland
Telephone: (011)(48)(22) 27-371
Fax: (011)(48)(22) 27-371*

For the Katowice Center, contact:
Slawomir Pasierb
Director of Research and Development
Telephone: (011)(48)(832) 57-1197
Fax: (011)(48)(832) 57-1197*

Polish Sources of Financing:

* Polish-Foreign Ecological Foundation (POL-EKO). "Pol-Eko," a private ecological foundation established in 1989, is mandated to plan cooperative ventures with foreign partners in the production and installation of pollution control and monitoring equipment, recovery of ecologically devastated land and water, and establishment of an ecological consulting service.

Mr. Zbigniew Peszko
Polish-Foreign Ecological Foundation (POL-EKO)
ul. Madalinskiego 39/43
02-544 Warsaw
Poland
Telephone: (011)(48)(22) 49-23-62

* Polish Development Bank. This Development Bank, established in September, 1990, is financially supported by the European Community to support privatization. Banking services are available to foreign and local investors and entrepreneurs in zloty and hard currencies.

* Fund for Environmental Protection and Water Management. This fund is financed by the collection of fines and penalties from enterprises that exceed pollution limits. A portion of the revenues are placed in a central fund, which is available for investment in pollution control devices that limit sulfur dioxide emissions. The remainder is channelled to Poland's regional territorial units or voivodeships. Up to 50 percent of zloty investment credits from the funds is available to those enterprises that invest in pollution control equipment. Contact the Ministry of Environmental Protection, Natural Resources, and Forestry. See Polish Contacts in Section V.

Multilateral Sources of Financing:

* Regional Environmental Center. The Regional Environment Center in Budapest, Hungary was established in September, 1990. The main objectives of the Center are to collect and disseminate data, develop institutional capabilities and outreach programs, promote public awareness and develop educational curricula, and serve as a clearinghouse on environmental information. The Center will accept proposals from individuals, non-governmental organizations, corporations, and governments. Proposals should state the specific role of the Center and to environmental improvement in Eastern Europe.

Proposals should be submitted in five (5) copies to the following address.

Mr. Stephen Wassersug
Program Manager
Regional Environmental Center
Miklos ter 1
Budapest, Hungary
Telephone: (011)(36)(1) 168-6284
Fax: (011)(36)(1) 168-7851

* World Bank. The World Bank or International Bank for Reconstruction and Development (IBRD) is a multilateral development bank that provides credits and loans to developing countries for large infrastructure projects. Loans are granted to the member governments that, in turn, solicit bids from companies around the globe for the procurement of consulting services or equipment necessary to execute projects. The World Bank undertakes a healthy number of projects in the environment/energy sectors, especially in the Eastern Europe region.

U.S. companies are encouraged to track and bid on projects in these sectors. For further information on current or forthcoming World Bank projects, consult the World Bank monthly publication "Monthly Operational Summary." This summary is available for your perusal at the U.S. Department of Commerce's International Major Projects Reference Room 2203 or by subscription to "International Business Opportunities Service." To order subscriptions, call (201) 225-2165.

For further information on World Bank activities relating to Poland, contact Richard Ackermann at (202) 473-2606 or contact David Craig at (202) 473-2589.

* International Finance Corporation (IFC). The IFC is an affiliate of the World Bank that lends to the private sector by providing debt and equity assistance. The IFC also coordinates financing from commercial banks, export credit agencies, and other institutions, as well as arranges for debt for equity swaps and the syndication of large loans.

Mr. Martyn Riddle
Chief, Environmental Unit
International Finance Corporation
1818 H Street, NW
Room I-10145
Washington, DC 20244
Telephone: (202) 473-0661

* European Bank for Reconstruction and Development (EBRD). Inaugurated April of 1991, the EBRD is a multilateral finance institution established to promote private and entrepreneurial initiative through loans and equity investments. Environmental and energy projects are a special focus of its lending activities.

For general inquiries, contact its head office in London:

Ms. Carol Wafa or Ms. Laura Nicholson
European Bank for Reconstruction and Development
122 Leaden Hall
London EC3V 4QL
England
Telephone: (011)(44)(71) 338-6000

For inquiries specifically for Poland on Merchant Banking, contact Mr. Bill McCauley and for Development Banking in the environmental and energy sectors, contact Mr. Piotr Wilczynski at the above telephone number.

* Commerce Business Daily: "Commerce Business Daily" is a daily publication that lists trade and investment contract opportunities and solicitations for bids by U.S. Government agencies, multilateral organizations, and foreign governments. Companies are encouraged to track these listings routinely. You may obtain copies through the Government Printing Office (GPO) by calling (202) 783-3238 or order a subscription by contacting:

Commerce Business Daily
11300 Rockville Pike, Suite 1100
Rockville, MD 20852-3030
Tel: (301) 961-8777

-or-

P.O. Box 5999
Chicago, IL 60680
Tel: (312) 353-2950

V. ENVIRONMENTAL CONTACTS IN POLAND

Ministry of Environmental Protection, Natural Resources and Forestry (MOSZNL)
00-920 Warsaw
ul. Wawlska 52/54
Contact: Minister Bronislaw Kaminski
Telephone: (011)(48)(22) 25-00-01 or 25-40-01
Telex: (011)(48)(22) 81-28-16

Foreign Investment Agency
00-950 Warsaw
Plac Trzech Krzyzy 5
Contact: Zbigniew Piatrovsky, President and Hubert Janiszewski, Vice President
Telephone: (011)(48)(22) 29-34-776
Telex: (011)(48)(22) 81-45-01

Ministry of Foreign Economic Relations
00-507 Warsaw
pl. Trzech Krzyzy 3
Contact: Minister Marcin Swiecicki
Telephone: (011)(48)(22) 69-35-000
Telex: (011)(48)(22) 81-45-01 MHZ PL
Fax: (011)(48)(22) 29-06-17 or 28-68-08

Polish Chamber of Foreign Trade
00-950 Warsaw
ul. Trebacka 4
Contact: Tadeusz Zylkowski
Telephone: (011)(48)(22) 26-02-21
Telex: (011)(48)(22) 81-43-61 PIHZ PL

Polish Academy of Sciences
Institute of Geography & Spatial Sciences
Plac Kultury i Nauki #2215
00-927 Warsaw
Contact: Dr. Andrzej Kassenberg, Director
Telephone: (011)(48)(22) 20-02-11, ext. 4123

Polish Ecological Club
Krakow 31-031
ul. Garbarska 9
Contact: Dr. Maria Guminska
Telephone: (011)(48)(12) 22-21-11 or 22-22-74

U.S. Commercial Officer
U.S. Trade Center/American Embassy Poland (WAW)
APO, NY 09213
Contact: Mr. Edgar Fulton, Sr. Commercial Officer
Telephone: (011)(48)(22) 21-45-15
Fax: (011)(48)(22) 21-63-27
Telex: 813934 USTDO PL

VI. OTHER ENVIRONMENTAL CONTACTS

Business Opportunities:
Eastern Europe Business Information Center
U.S. Department of Commerce
Room 6043
14th & Constitution Avenue, NW
Washington, DC 20230
Contact: Lydia La Ferla
Telephone: (202) 377-2645

Energy Information:
U.S. Department of Energy
Room 4G-067
1000 Independence Avenue, SW
Washington, DC 20585
Contact: Dennise Swink
Telephone: (202) 586-9680

Environmental Regulation:
U.S. Environmental Protection Agency
401 M Street, NW
Washington, DC 20460
Contact: Amy Evans
Telephone: (202) 382-4897

Regional Environmental Center:
1035 Budapest
Miklos Ter 1
Contact: Mr. Stephen Wassersug
Telephone: (011)(36)(1) 168-6284 or 168-8685
Fax: (011)(36)(1) 168-7851

EPA Liaison Contact: Dan Thompson
Telephone: (202) 382-5224

Polish Commercial Office:
Office of the Commercial Counselor to the Polish Embassy
820 Second Avenue, 17th Floor
New York, NY 10017
Contact: Mr. Jerzy Kapuscinski
Telephone: (212) 270-5300

Polish-U.S. Economic Council
U.S. Chamber of Commerce
1615 H Street, NW
Washington, DC 20062
Contact: Mr. Beurt SerVaas Chairman and CEO, Curtis International Ltd.

Mr. Thaddeus Kopinski
Executive Director
Telephone: (202) 463-5482
Fax: (202) 463-3114
Telex: RCA 248302

Polish Ecological Club
U.S. Representative
Humphrey Institute of Public Affairs
Room 152 Humphrey Center
Minneapolis, MN 55455
Contact: Professor Zbigniew Bochniarz
Telephone: (612) 625-5527

U.N. Environmental Programme
1889 F Street, NW
Washington, DC 20006
Contact: Ms. Joan Martin Brown
Telephone: (202) 289-8456

U.S. Address for U.N. Environmental Programme:
2 UN Plaza
Room DC2/803
New York, NY 10017
Contact: Dr. Noel Brown
Telephone: (212) 963-8138
Fax: (212) 963-7341

U.N. Economic Commission for Europe
Environment & Human Settlements Division
Palais des Nations, 1211
Geneva 10
Switzerland
Contact: Mr. Peter Sand
Telephone: (011)(41)(22) 734-6011

VII. ENVIRONMENTAL TRADE PROMOTIONAL EVENTS CALENDAR FOR EASTERN EUROPE: 1991-1992

1991

September 1991: **Conference on "Challenge of the Environment: New Technologies, New Markets."** Taking technologies out of National labs into the private sector. Sponsored by the U.S. Department of Commerce's Technology Administration on September 5-6, 1991 at the Sheraton Reston, Reston, VA. Contact Evelyn Fields at 202/377-8103.

October 5-19, 1991 **California IOGA Environmental Products and Technologies Trade Mission** to Prague, Czechoslovakia; Budapest, Hungary; Dresden, Stuttgart, and Frankfurt, Germany. Sponsored by the World Trade Center Association of Orange County. Contact Jay De Long, Director of Development at 714/752-9822.

October 1991 - September 1992 **Pollution Control Catalogue Show** to Belgrade, Yugoslavia;
Krakow and Warsaw, Poland; Prague, Czechoslovakia; Budapest, Hungary, Sofia, Bulgaria, and Ankara, and Istanbul. Sponsored by U.S. Department of Commerce. Contact Sandra Collazo, Project Manager at 202/377-0871 or fax at 202/377-0178.

November 13-27, 1991 **Air Pollution Control Technologies Trade Mission** to Poland, Romania, and Bulgaria. November 13-27, 1991. Sponsored by U.S. Department of Commerce. Contact Tyrena Holley, Project Officer at 202/377-3509 or fax at 202/377-8304.

1992

November 23-26, 1992	**POLERO International Ecological Fair.** Poznan, Poland. Features environmental control and measuring equipment, waste-free technologies and biotechnologies, energy generation methods, and ecological education. Exhibitions are accompanied by symposia, seminars, and scientific and technical conferences. Contact Wydzial Handlowy, Poznan International Fair, 60-734 Poznan, ul. Glogowska 14, Poland. Telephone (011)(0)(61) 66-37-74 or fax (011)(0)(61) 66-58-27.
April 1992	**USA Tech Solo Trade Fair** in Bratislava, Czechoslovakia, which includes environmental technologies, renewable energy technology, pollution control equipment. Featuring a separate pavilion for the exhibit of U.S. clean coal technology and panel discussions to provide potential buyers with technical information. Contact Damon Greer at 202/377-0564.

APPENDIX H - ENVIRONMENTAL SUMMARY: HUNGARY

This appendix was made available by TMS Management Consulting, 6 Mayflower Road, Framingham, MA 01701. Dated January 1992. The author is Kenneth J. Macek, Ph.D.

The project support was provided by the U.S. Agency for International Development under the contract number ANE-0178-C-00-0041-00.

EXECUTIVE SUMMARY

Four decades of a Communist command economy in Hungary has produced an environment that suffered serious environmental insults. Hungary is developing a comprehensive strategy for restoration and protection of its environment. This strategy includes both legislative and economic mechanisms necessary for its successful implementation. The final strategic component necessary for the successful implementation of Hungary's emerging strategy is the development of thriving domestic environmental services and products sectors.

The market for such goods and services had not exhibited measurable growth over the past five years. Growth rate forecasts are in double digits for the foreseeable future. Most of the anticipated expenditures will be in the areas of hazardous waste cleanup and air and water pollution control. Whether or not viable growing markets actually emerge is mostly a function of the political will in Hungary to incorporate environmental considerations into the rapid legal and economic changes occurring during the social and market transformation of the country.

Many of the technical skills required to develop a diverse environmental services sector currently exist in Hungary. There are currently some 100 environmental enterprises in the CSFR offering a variety of design, engineering and consulting services. There is however a wide disparity between the commercial resources required and those available. Given appropriate technical training and management resources, Hungary should be able to satisfy the environmental services needs of its economy and society.

The variety of environmental products currently produced in the Hungary is extremely limited. There appears to be a far smaller number of suppliers of environmental products than services. Competitiveness of products produced in Hungary suffers from both lower technology, due to past isolation from the West, and higher costs due to management and resource inefficiencies. However, given access to technology and proper training, the technical expertise available in the manufacturing sector in Hungary appears to have the capability to produce many of the products required in the environmental sector.

INTRODUCTION

During the 40 years following the end of the war in Europe, the economic structure of Hungary changed radically under the Communist regime. A period of rapid industrialization, with emphasis on heavy industries (mining, metallurgy, chemicals), occurred without regard to the environment. The command economy's emphasis on fulfillment of "the plan" encouraged industrial progress at the expense of resource efficiencies.

This was particularly true in the energy sector where dependence upon untreated, low-quality domestic coal was a major contributor to environmental degradation. The Communist government in Hungary paid lip service to the issue of environmental protection, while priorities remained production and quotas. This resulted in major environmental insults to the country, though possibly not as egregious as those observed in Poland and other neighboring countries.

At the time of the overthrow of the Communist regime, the people and their newly elected leaders in Hungary placed environmental issues on a par with the country's other serious issues. Nothing illustrated this concern for the environment more than the public outcry against the country's involvement in Gabcikovo-Nagymaros Dam complex, designed to provide hydroelectric power to Austria, Hungary and Czechoslovakia. As a result of this public outcry, the new government delayed, and then cancelled its participation in the project, purportedly because of environmental concerns.

However, following the initial environmental euphoria, the realities of transforming even the most "westernized" of the command economies to a market economy began to dominate the government's policies and programs. Once again, economic issues appear the paramount concern, possibly at the expense of the environment. This posture is not unexpected. To the people, the quality of the environment becomes significant only above a certain level of the quality of life. Thus, it is incumbent upon the government of Hungary to identify environmental protection as an economic issue, having long range economic advantages. Environmental policy can be efficient only if the elements of environmental regulation are incorporated into the laws governing development and the transformation of the economy.

Organizational, business development, financial, marketing and management training assistance are also essential to the development of a strong domestic environmental sector. Some specialized technical training (e.g. waste minimization, pollution prevention, energy efficiency, risk assessment, etc.) are required to strengthen the domestic environmental services sector in Hungary. Access to western components, technologies, and manufacturing techniques, sufficient investment, and direct entrepreneurial assistance are required to stimulate the development of an environmental products sector.

Today, Hungary is taking nominal steps to address its myriad environmental problems. It has established a Ministry of Environment and Regional Policy (MERP). The staff at MERP is working diligently to develop an integrated environmental strategy, based on least cost approaches that emphasize economic incentives and institutional modifications rather than investments. Most significantly, this strategy includes not only environmental legislation and regulations, but also the appropriate domestic economic mechanisms required to carry out its environmental programs. The transition from a posture where the government owns the "industrial sector" and is responsible for pollution, to one where it regulates the industrial sector, of necessity will require a significant period of time.

The final strategic component necessary to restore and protect Hungary's environment is a thriving domestic environmental business sector. As expected, in a economic system that gave environmental protection a low priority these sectors are extremely limited. Today, as Hungary moves away from a command toward a market economy, viable environmental goods and services sectors are beginning to slowly emerge. However, Hungary requires technologies, technical assistance and training from the West, and exposure to market and business management principles to nurture and sustain these essential business sectors.

The objectives of this report are to characterize both the environmental market potential and the domestic capacity to satisfy those markets; and to identify opportunities for foreign assistance and investment to strengthen the domestic environmental goods and services sectors in Hungary.

ENVIRONMENTAL PRIORITIES IN HUNGARY

Estimates to restore Hungary's environment approach $50-100 billion. The government is developing a strategy that includes investing approximately $10 billion, over the next twenty years, in infrastructure projects to address some of these environmental concerns. The private sector will be expected to make even greater investments during that period to meet environmental pollution control and abatement.

The priorities identified below indicate that Hungary intends to combat pollution at the source through changes in resource efficiency and technology. The environmental priorities of the new Ministry of Environment and Regional Planning (MERP) are focused on the abatement of air pollution, neutralization of hazardous wastes, and development of sewage treatment:

AIR PROTECTION

A. Reduction of emissions of sulfur dioxide, nitrogen oxide and other pollutants from power plants, chemical and other basic industrial plants, and vehicular traffic by:

- desulfurization of hard coal
- utilization of fluidized bed boilers
- gasification of coal with gas-steam turbines
- desulfurization of gases
- introduction of catalytic converters and use of lead-free gasoline
- development of alternative energy sources
- upgrading air quality monitoring systems

WASTE MANAGEMENT

A. Hazardous waste management, including:

- construction of incineration plants for toxic and hazardous wastes
- economic use of pyrite waste resulting from the desulfurization of hard coal
- economic use of "red mud" from aluminum production
- recovery of metals from galvanic wastes.

B. Municipal waste management, including:

- production of equipment for selective collection, transportation, processing and storage
- segregation of municipal waste with recovery and recycling of raw materials
- development of composting facilities

WATER PROTECTION

A. Development of river basin management systems and urban potable water supply;

B. Installation of efficient municipal wastewater treatment plants with full processing of sludge;

C. Protection and monitoring of changes in the karst system;

D. Manufacture of equipment for wastewater treatment plants, such as aerators, pumps, sludge dewatering systems, automatic process control systems;

E. Monitoring of industrial wastewater discharges.

As Hungary implements its environmental programs, it will require a wide variety of environmental goods and services. Most of these products and services are not available in the domestic environmental sectors. When the demand for the required goods and services exists, it should produce sustainable environmental markets

Today, Hungary faces two major challenges. First, it must fully develop those domestic skills, resources and assets which can lead to the development of competitive, least cost environmental solutions. Second it must access the Western technology, expertise and skills it lacks in order to fulfill its long-term environmental objectives.

DOMESTIC ENVIRONMENTAL MARKETS

STRUCTURE

Demand for environmental goods and services is largely generated by public concern about the environment and by governments enacting and enforcing environmental legislation. When analyzing the structure and potential of environmental markets in Eastern European countries, the relevance of the conventional "Western" market paradigm (FIGURE 1) to these markets is questionable, because most components are absent.

The only component of this paradigm currently in place in Hungary is the potential customer base and its plethora of environmental needs. However, this component alone does not constitute a dynamic environmental market. The extent to which the other necessary market mechanisms and forcing functions develop in Hungary will determine the size and timing of its environmental markets. Each paradigm component must be carefully considered when assessing the current and future environmental markets in Hungary.

The domestic environmental markets in Hungary mirror the state of the chaotic and rapid transition occurring throughout the economy and country. Characterizations of these markets are extremely difficult due to lack of adequate data. Unlike Poland, the United Nations Industrial Development Organization has not been able to establish a reliable data base on this sector. Unlike Czechoslovakia, there is no private sector market research group that has described the environmental market sectors in detail.

Only limited data exists as a result of a single effort by MERP to identify enterprises active in the environmental business sectors. Nonetheless, an environmental market must exist in Hungary. The key issues are estimating its current size and forecasting its future growth. To do this, it is necessary to assess all of the market paradigm components as they exist in Hungary today.

Throughout Eastern Europe, the institutional capability to promulgate, implement and enforce environmental regulations is now being developed. However, at this time, Hungary appears to be lagging behind its neighbors in this regard.

It has not yet developed a complete and comprehensive environmental strategy. As a result, the environmental legislation and economic funding mechanisms required to structure and fund environmental programs appear least developed there.

Finally, the political will to give environmental issues a priority, equal to other issues associated with economic transformation, represents the most important variable in development of domestic environmental markets in the region. To this observer, political will and leadership appears to be weaker in Hungary than in Poland and Czechoslovakia.

ACTIVITY

During the period 1987-91, reported federal environmental expenditures in Hungary were relatively flat and averaged some SISo million annually. The annual corporate investments for environmental protection during the same period were about the same level as that of the government. This is somewhat surprising, as the government traditionally accounted for most of the environmental investment in other Communist countries in Eastern Europe. It is difficult to

Figure 1
Schematic Diagram of the Conventional Western Paradigm for Environmental Markets

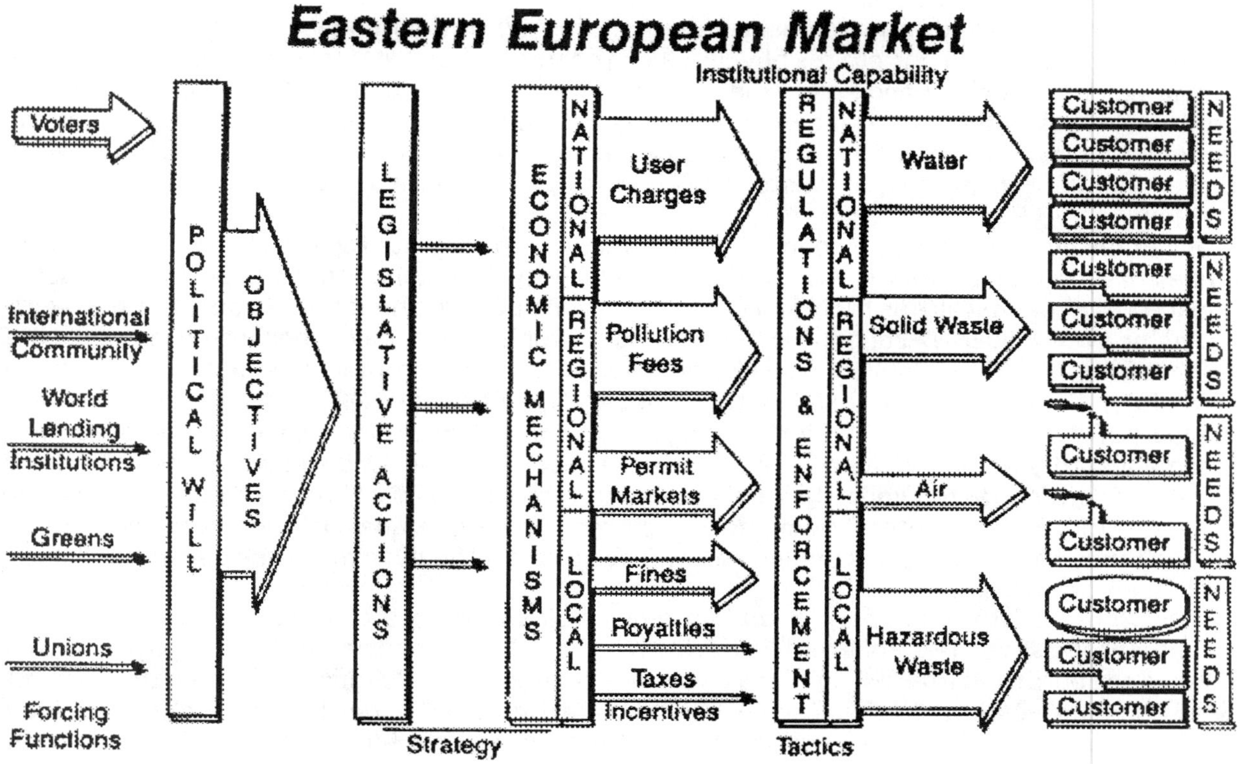

imagine that the total investment in environmental goods and services in Hungary exceeded $300 Million during 1991. Much of this investment subsidized end-of-the-pipe technologies.

The Ministry of Finance estimates that, during the period 1992-96, the government of Hungary will spend approximately $500-700 million annually to achieve short-term environmental goals. The government forecasts that the total investment in environmental infrastructure for water supply and sewage treatment alone will exceed $1 Billion during this period. The expected increase in the size of the private sector, as well as the potential increased environmental regulation and enforcement should cause investments by the private sector to increase more rapidly than those of the public sector. However, whether or not the government has the political will to deal with economic and environmental issues simultaneously is presently unclear.

FINANCING

A major concern regarding the future environmental products and services sectors in Hungary is the government's ability to finance its goals. Parallel with the development of the legal and technical means of environmental protection, it is necessary for Hungary to elaborate an economic regulatory system encouraging preventive environmental protection, as well as to incorporate environmental protection requirements into the general economic system.

Several international organizations such as the World Bank, the European Bank for Reconstruction and Development, and the EC have provided tens of millions of dollars in grants and loans to Hungary for environmental programs. The United States will provide Hungary with approximately $10-20 Million in environmental aid over the next several years. Other bilateral agreements, requiring matching funds, contribute significantly to these efforts. The major objectives of this international aid are (a) institutional strengthening; (b) technology transfer; and (c) installation of pollution monitoring systems.

However, in view of the environmental expenditures required in Hungary over the next decade, the role of international financial aid is obvious. It can only serve as a catalyst to the domestic market for environmental products and services in Hungary. The major source of funds must come from the limited resources within the Hungarian economy. Presently, the major source of funding for environmental investment is a system of fines covering seven fields: air pollution (20%), sewage and drainage (60%), hazardous waste (16%), noise protection, nature conservation and land protection.

Pollution levels are set by a county board or regional directorate for each facility. In the past, the fines have been inadequate. Also, if the local authority determined that the polluting industry exhausted all economically or technically feasible means of reducing its emissions, the fine were reduced, or even waived. Recently, the fines for water, for example, have been increased to such a level that industries are seriously considering investment in control technologies.

In effect, Hungary proposes to adopt the *Polluter Pays Principle* (PPP). Under this principle, individual polluters are responsible for meeting environmental requirements and for compensation for damages to the environment. There are, however, several limitations to the PPP. It is enforcement intensive, requiring the government to monitor the discharges and expenditures of polluters, or potential polluters. Some polluters do not have the financial resources required to fulfill their individual obligations under the law. Lastly, it is difficult to make the current polluters pay for their past neglect.

Thus, both equity and financial realism require public sources of funds to address certain environmental issues. The Polluters Pay Principle (PsPP) relieves the general taxpayer of this additional tax burden. With PsPP, charges to current polluters provide funds to finance that portion of environmental restoration that cannot be adequately financed under the PPP. Entities such as national and regional funds for environmental protection receive fees, taxes and/or fines from current producers (polluters).

The structure of solid waste disposal charges will be governed by principles similar to those pertaining to water resources management. User charges will be utilized to support investment in and operation of collective facilities. In addition, product charges to encourage recycling are also being considered.

In air quality control, usually there are no activities which require joint treatment activities. Thus, each "user" of air can be made individually responsible for necessary pollution abatement activities. Also, air quality control programs may utilize marketable permits. In addition, the government envisions levying product charges on environmentally harmful goods. A proposed tax on gasoline is expected to generate up $100 million annually.

Other mechanisms such as soft loans, tax allowances, reduced import tariffs, etc. will be utilized where appropriate to accomplish the government's goals relating to environmental restoration and protection.

STRUCTURE

The environmental services sector in Hungary is in its infancy. Today there are approximately 100 firms, mostly small (i.e. 5-50 employees) providing environmental services. These firms emanate primarily from burgeoning entrepreneurial activity and, to lesser extent from fragmentation of the larger government organizations now being privatized. There is a handful of large firms (i.e. hundreds of employees), some of which have demonstrated that they can successfully compete in international markets. Today, the environmental services sector consists mostly of engineering and consulting firms involved in the treatment and disposal of hazardous and municipal wastes. The next largest area of activity is engineering and design of wastewater treatment systems, both industrial and municipal.

The environmental products sector is currently extremely limited. This may be due to the lack of significant light manufacturing capabilities in the industrial sector. It may also result from the fact that Hungary was isolated from the technology developments occurring in this sector in the west during the last decade. However, this dearth in the domestic environmental products sector may also reflect the lack of demand, both past and current, for such products in Hungary.

Air filtration devices for dust collection and packaged sewage treatment plants have dominated the product sector. In recent years, some waste treatment technologies for industrial wastewater are appearing in this marketplace.

Appendix I identifies firms providing environmental products and services and the major market segments served by each.

ENVIRONMENTAL SERVICES

OVERVIEW

The environmental services sector has only recently emerged in Hungary's fledgling market economy. Hungary requires significant investment in training, technology and time before its environmental services sector is comparable to, and competitive with, Western counterparts. A significant services sector will not emerge until significant market demand is created, a phenomenon whose arrival is not clearly evident at this time.

Hungary has environmental engineers qualified to design and construct waste treatment facilities, water and wastewater treatment facilities, and crude (by Western standards) air pollution control systems. This talent pool exists in state owned specialized engineering organizations. They focus on areas such as waste neutralization, sanitary engineering, steel industry wastewater and air particulates control, etc. Most are, or will be, privatized, resulting in fragmentation into smaller, specialized "consulting" enterprises.

Also, Hungary's excellent academic and research institutions support a cadre of environmental scientists (ecologists, water quality experts, ecotoxicologists, hydrogeologists, public health experts, etc.) Most of these scientists have mainly been involved in basic research, and may find the transition to applied science (i.e. consulting) difficult. Also, (like their engineering counterparts), they lack the organizational, management, marketing and financial skills required to operate an enterprise in a market economy.

However, the technical skills required to develop a diverse environmental services sector in Hungary do currently exist. Given organizational and business development assistance, the introduction of financial, marketing and management skills, specialized technical skills training (e.g. environmental auditing, waste minimization, pollution prevention, hazard and risk assessment), and access to key western technologies, Hungary should be capable of satisfying the environmental consulting services needs of its economy and society.

There are a few large firms that offer general consulting services within the broad sphere of environmental protection. These include MELYEPITESI TERVEZO VALLALAT, ANYAGVIZSGALO ES GEPIRARI MINOSEGELLENORZO INTEZET, and SZIKKTI (Budapest), BIOTECH (Domsod), COMENVIRON (Debrecen) and TBVFI (Tatbanya).

By far the largest number of consultants are active in the treatment and disposal of hazardous wastes. The amount of waste produced in Hungary is controversial, as most estimates are provided by the waste producers. The government claims that 50% of the approximately 100 million tons of industrial waste produced annually is recycled, and that only 5% is considered hazardous. The majority of this hazardous waste consists of red mud (the waste from aluminum production).

Other major segments of the environmental business sectors apparently have not yet attracted large numbers of firms. The wastewater treatment and the water supply sectors appear to have the largest numbers of suppliers, after the waste management segment. Surprisingly, there apparently are not a large number of domestic firms involved in serving the air pollution control or analytical services segments.

WASTE MANAGEMENT

More than half of all environmental services firms identified indicted that they offered services in the area of hazardous wastes management. The largest firms in this segment appear to be CSONGRAD MEGYEI TELEPULESTISZTASAGI (Szeged), ENVIRON, GEOHIDRO and KOZPONTI BANYASATI FEJLESZTO INTEZET (Budapest), FLORASCA (Sopron), SWIETELSKY-DUVIEP (Veszprem), HEVES MEGYI TELEPULES-TISZTASAGI (Eger) and GYOGYSZERIPARI EGETOMU XOZOS VALLALAT (Dorog).

Several firms offer very specialized waste handling services. GYOGYSZERIPARI EGETOMU KOZOS (Dorog) specializes in the transportation of hazardous wastes. ALKOTO (Budapest) and BIOFER (Nyiregyhaza) specialize in the clean-up and disposal of petroleum wastes. METALLOGLOBUS (Budapest) specializes in handling wastes from processing of non-ferrous metals, while HYDROCOR (Budapest) specializes in handling slag from the paints and pigments industry. GYAPAROS KORNYEZETVEDELMI MUNKAKOZOSSEG (Bekescsaba) and VEGYI ES R0BBANOANYAGIPARI FELUGYELET (Budapest) specialize in solvent regeneration.

GRABOCENTER (Gyor), PG ACELSZERKEZETGYARTO ES FORGOMALZO (Biharkerestes), and PRODINFORM and PROMETHEUS (Budapest) offer services in the incineration of industrial wastes. KEMIKONTROL and AQUA-VITA (Budapest) appear to be the only firms offering bioremediation services.

There are currently no incentives to recycle household waste nor to separate different types of waste. Thus less than 10% of municipal wastes are reused or recycled, while the remainder is dumped in landfills (including household wastes considered hazardous). Hungary is already experiencing a critical shortage of suitable dump sites. The only waste incinerator for municipal wastes is in the city of Budapest.

The leading services firms in the municipal waste management segment appear to be FLORASCA (Sopron), SWIETELSKY-DUVIEP (Veszperem) and FOVAROSI KOZTERULETFENNTARTO VALLALAT and KOZPONTI BANTASZATI FEJLESTO INTEZET (Budapest). Some twenty other firms indicate that they provide services to this market segment.

WATER MANAGEMENT & PROTECTION

The field of water management and pollution control is estimated to be the largest segment of the environmental sector. The bulk of Hungary's water resources come from within the international water basin. Thus the availability and quality of water in Hungary is determined largely by its neighbors. Roughly two-thirds of the available water is surface water, and one-third is groundwater. The growth of water demand by industry, agriculture and municipalities has grown dramatically in recent years. Today almost 75% of this demand is from industry, 15% from agriculture, and less than 10% from municipalities.

In order to appreciate this segment, one must understand an environmental aspect, which may be unique in the region, Hungary's potable water supply system. The bulk of the country's surface water is perceived as already polluted. Yet almost 50% of the country's drinking water is ultimately derived from these surface water sources through a process called "bank filtration." The Danube provides almost 90% of bank filtered water, serving as the primary source to the city of Budapest.

AIR POLLUTION

The most important contributors to deteriorating air quality are coal/lignite fired power plants and household furnaces. The primary air pollutants are particulates (120,000 tons per year), sulfur dioxide from lignite, and nitrogen compounds from vehicular traffic (one-third of the vehicles have two stroke engines).

Almost one-third of the firms identified in the environmental services sector indicated that they are active in the air quality segment. The major diversified firms in the air quality segment appear to be ANYAGVIZSGALO ES GEPIRARI MINOSEGELLENORZO INTEZET, KOHASZTI GYAREPITO VALLALAT and KORNYEZETTECHNIKA MUSZAKI ES SZOLGALTATO (Budapest), DIMED MUSZERTECHNIKAI (Miskolc) and FUTOBER (Batonyterenye).

Other more specialized service firms in the air quality segment include AMBICIO and AUTOTRIB (Budapest) which focus catalytic exhaust elimination; and ENIKOM, MECHATRONIKA, and IINOTRADE (Budapest), SZEGEDI VAROSGAZDALKODASI (Szeged) and DATAQUA ELEKTRIKAI (Balatonalmadi) which focus on air emissions measurements.

OTHER ENVIRONMENTAL SERVICES

There are a surprising number of firms involved in the noise pollution abatement segment (presumably because there exists regulations and penalties associated with noise pollution). These firms include AET SZIGETELO ES SZOLGALTATO, AUTOIPARI KUTATO ES FEJLESZTO VALLALAT, ENIKOM GLOBUS-SYNCRONIC, ATRI EPITOIPARI and OLAJTERV (Budapest). AKTIV (Kapolnasnyek) specializes in leak detection from underground storage tanks. ATLAS (Budapest) focuses on remediation of contaminated soil. BIOFER (Nyiregihaza) specializes in asbestos removal. BOLYAI KATONIA FOISKOLA and TRADEXPERT (Budapest) provide general analytical services.

By far the greatest number of domestic environmental equipment suppliers serve the water and wastewater segments. The government's environmental strategy includes massive investment in municipal and industrial wastewater treatment to provide cities and industry with water of sufficient quality for use. In addition to packaged treatment plants, other equipment priorities include sludge dewatering and control equipment, and systems for wastewater treatment plants.

The major suppliers to these segments appear to be BUDAPESTI KOOLAJIPARI GEPGYAR (Budapest), which supplies wastewater filtration systems and anti-emulsion agents, BUDAPESTI VEGYIPARI GEPGYAR (Budapest), which supplies filter presses and fermentation tanks, MAGYAROVARI TIMFOLD ES MUKORUNDGYAR (Mosongyarovar), which provides water treatment chemicals, and MELYEPITESI TERVEZO VALLALT (Budapest), which provides monitoring equipment.

Other specialized suppliers include METEFEM (Budapest) which supplies dryers, shakers and mixers, KORTE (Dunaharaszti) which provides chemical feeders, HYDROFER IPARI (Szentes) which provides packaged water and wastewater treatment systems, and HIDROPLASZTIK (Nagykanizsa) which provides water softeners.

AIR POLLUTION

This segment is extremely small when one considers the magnitude of the air pollution problems facing Hungary. It is obvious that significant technology transfer is necessary to enable Hungary to deal with these problems. The number of suppliers and the products supplied appear grossly inadequate at the present time.

INNOTRADE (Budapest) and FUTOBER (Batonyterenye) supply different types of filters for dust control. AMBICIO and AUTOTRIB (Budapest) provide equipment and chemical additives for reducing emissions from internal combustion engines. DIMED MUSZERTECHNIKAI (Miskolc) provides diagnostic instrumentation for measuring air emissions.

CONCLUSIONS

The events in Eastern Europe, over the last few decades, are testimony that serious environmental problems, *per se*, do not create environmental markets. A chain of events, involving the government, the people, and various segments of the private sector, must occur, on a national scale, before viable environmental markets emerge.

Capitalists like to think that efficient market forces are sufficient to internalize all true costs, even environmental costs. However, care for the environment is not a by-product of market economies. Western industrialized societies have discovered that market forces, in and of themselves, are not sufficient to create environmental markets. on the contrary, every western industrialized society has had to rely on a hybrid regulatory approach consisting of command and control elements to establish objectives, and market mechanisms to provide least-cost solutions. Hungary and other emerging democracies of Eastern Europe must do the same.

In view of the myriad economic problems facing countries in transition from a command to market economy, the inclination to place environmental issues on the "back burner" is understandable. However, until the political will to internalize environmental costs exists within a government, there is unlikely to be significant progress in addressing the environmental ills which plague countries in economic transition. In view of the current lack of legislative reform and institutional development in Hungary, the near term development of a dynamic, domestic environmental products and services sectors is highly suspect.

The private environmental business sector in Hungary is only now emerging. There are some areas of expertise, and many essential environmental technologies, systems and products that are not yet available domestically. However, the potential to provide the required services is great. On the other hand, the potential for domestic development of the required environmental products and technologies appears to be quite limited. This sector will require significant investment and technical support.

Appendix I
Domestic Environmental Products and Services Suppliers

Company		Air	Water	Waste Water Treatment		Waste Treatment		Analytical	
(Location)	Type*	Pollution	Supply	Industrial	Municipal	Industrial	Municipal	Services	Other
AET (Budapest)	S								xx
AGMI (Budapest)	S	xx				xx		xx	
AKKA (Budapest)	S	xx	xx	xx					
AKITIV (Kapolnasnyek)	S		xx						
ALFA (Salgotarjan)	P		xx						
ALISCA PATENT (Szekszard)	P/S		xx			xx	xx	xx	
ALKOTO (Budapest)	P/S					xx			
AMBICIO (Budapest)	P/S	xx		xx					
ANNYAGVIZSGALO ES GEPIPARI MINOSEGELLENORZO INTEXET (Budapest)	S	xx	xx			xx	xx		

[P—Provides products; S—Provides services; P/S—Provides both products and services]

Appendix I (*Continued*)
Domestic Environmental Products and Services Suppliers

Company (Location)	Type*	Air Pollution	Water Supply	Waste Water Treatment Industrial	Waste Water Treatment Municipal	Waste Treatment Industrial	Waste Treatment Municipal	Analytical Services	Other
AQUARIUS (Budapest)	S		xx						
AQUA-VIVA (Budapest)	S			xx					
BUDAPESTI KOOLAJIPARI GEPGYAR (Budapest)	P/S	xx		xx	xx	xx			
BUDAPESTI VEGYIPARI GEPGYAR (Budapest)	P/S		xx	xx					
BVM (Budapest)	P		xx	xx	xx				
BELAPATFALVI CEMENTES MESZIPARI (Belapatfalva)	S					xx			
CENTRISERV (Budapest)	S			xx	xx				
CHEMITECH (Dunaharaszti)	S		xx						
COMENVIORN (Debrecen)	S				xx	xx			
CSONGRAD MEGYEI TELEPULESTISZTASAGI (Mako)	S					xx	xx		
DATAQUA (Balatonalmadi)	P	xx						xx	
DEK (Miskolc)	S					xx			
DIMED (Miskolc)	P	xx						xx	
FOVAROSI KOZTERULETFENNTARTO (Budapest)	S						xx		
FOVAROSI TELEPULESTISZTASAGI (Budapest)	S				xx	xx			

[P—Provides products; S—Provides services; P/S—Provides both products and services]

Appendix I (*Continued*)
Domestic Environmental Products and Services Suppliers

Company (Location)	Type*	Air Pollution	Water Supply	Waste Water Treatment Industrial	Waste Water Treatment Municipal	Waste Treatment Industrial	Waste Treatment Municipal	Analytical Services	Other
FUTOBER (Batonyterenye)	P	xx							
GAMMA MUVEK (Budapest)	P	xx						xx	
GEOHIDRO (Budapest)	S					xx		xx	
GLOBUSO-SYNCHRONIC (Budapest)	S	xx							xx
GRABOCENTER (Gyor)	S					xx	xx		
GREENTECH (Budapest)	S					xx	xx		
GYOGYSZERIPARI EGETOMU MUNKAKOZOSSEG (Bekescsaba)	S					xx			
HAJDUSAGI KORNYEZETGAZDALKODASI (Debrecen)	S					xx			
KBFI-TRIASZ (Budapest)	S		xx					xx	
KEMIKONTROL KKT (Budapest)	P/S					xx		xx	
KEVTERV VKVM (Eger)	S				xx	xx		xx	
KEVITERV PLUSZ (Miskolc)	S					xx			
KISIPARI TERMELTETO (Budapest)	P	xx					xx		
KOHASZATI GYAREPITO (Budapest)	P/S	xx			xx				
KORNYEZETTECHNIKA MUSZAKI ES SZOLGALTATO (Budapest)	P/S	xx							

[P—Provides products; S—Provides services; P/S—Provides both products and services]

Appendix I (*Continued*)
Domestic Environmental Products and Services Suppliers

Company (Location)	Type*	Air Pollution	Water Supply	Waste Water Treatment Industrial	Waste Water Treatment Municipal	Waste Treatment Industrial	Waste Treatment Municipal	Analytical Services	Other
KORNYEZETVEDELMI KISSZOVETKEZET (Budapest)	S					xx			
KORTE (Dunaharaszti)	P/S			xx	xx				
KOZLEKEDESTUDOMANYI INTEZET (Budapest)	S	xx							
KOZPONTI BANYASZATI FEJLESZTO INTEZET (Budapest)	S					xx	xx		
NTSKI (Kaposvar)	S		xx			xx			
NYIREGYHAZI KOZTERULETFENNTARTO (Nyiregyhaza)	S					xx			
NYIRFA (Budapest)	S					xx		xx	
OBSERVATOR (Budapest)	S							xx	
OLAJTERV (Budapest)	S					xx			xx
OMIKRON (Mogyorod)	P							xx	
OMSZ (Budapest)	S	xx						xx	
PECSI VIZMU (Pecs)	S			xx					
PESTIDEKI GEPGYAR (Budapest)	P						xx		
PG ACELSZERKEZETGYARTO ES FORGOMALZO (Biharkeresztes)	P						xx		
PLUSZ (Budapest)	P/S	xx							

[P—Provides products; S—Provides services; P/S—Provides both products and services]

Appendix I (*Continued*)
Domestic Environmental Products and Services Suppliers

Company		Air	Water	Waste Water Treatment		Waste Treatment		Analytical	
(Location)	Type*	Pollution	Supply	Industrial	Municipal	Industrial	Municipal	Services	Other
PRESSKAN (Budapest)	S		xx		xx				
TEKS (Budapest)	S					xx			
TEMAFORG (Budapest)	P/S	xx	xx						
TRADEXPERT (Budapest)	P							xx	
TREDEX (Budapest)	S				xx				
TRIDEX (Budapest)	S	xx			xx	xx			
UNIFLEX (Budapest)	P					xx			
UNITEL (Budapest)	P/S		xx		xx			xx	
VASKUT (Budapest)	P/S	xx							
VAROSGAZDAKODASI (Turkeve)	S					xx			
VEKTT (Veszprem)	P/S	xx			xx				
VISITERV (Budapest)	S		xx	xx		xx	xx		
VITUKI (Budapest)	S			xx	xx	xx			
VRF (Budapest)	S					xx			

[P—Provides products; S—Provides services; P/S—Provides both products and services]

APPENDIX I - ENVIRONMENTAL SUMMARY: CZECHOSLOVAKIA

This appendix was made available by TMS Management Consulting, 6 Mayflower Road, Framingham, MA 01701. Dated January 1992. The author is Kenneth J. Macek, Ph.D.

The project support was provided by the U.S. Agency for International Development under the contract number ANE-0178-C-00-0041-00.

EXECUTIVE SUMMARY

Four decades of a Communist command economy has produced an environment in Czechoslovakia (CSFR) that is dismal by every standard. The CSFR is developing a comprehensive strategy for restoration and protection of its environment. This strategy includes both legislative and economic mechanisms necessary for its successful implementation. The final strategic component for the successful implementation of Czechoslovakia's strategy is the development of thriving domestic environmental services and products sectors.

The market for such goods and services had not exhibited measurable growth until 1991. It is forecast to continue growing at double digit rates for the foreseeable future. Most of the anticipated expenditures will be in the areas of air and water pollution control, and hazardous waste cleanup.

Many of the technical skills required to develop a diverse environmental services sector currently exist in the CSFR. There are currently some 150 environmental enterprises in the CSFR offering a wide variety of design, engineering and consulting services. There is however a wide disparity between the commercial resources available in the Czech Republic and those available in the Slovak Republic.

Given appropriate technical training and management resources, the CSFR should be able to satisfy the environmental services needs of its economy and society.

The variety of environmental products currently produced in the CSFR is limited, especially in the area of water purification. There appears to be a significantly smaller number of suppliers of environmental products than services. The disparity between the two republics in the number of product suppliers is again clearly evident. Competitiveness of products produced in Czechoslovakia suffers from both lower technology, due to past isolation from the West, and higher costs due to management and resource inefficiencies.

However, given access to technology and proper training, the technical expertise available in the manufacturing sector in Czechoslovakia appears to have the capability to produce many of the products required in the environmental sector.

Organizational, business development, financial, marketing and management training assistance are also essential to the development of a strong domestic environmental sector.

Some specialized technical training (e.g. waste minimization, pollution prevention, energy efficiency, risk assessment, etc.) are required to strengthen the domestic environmental services sector in the CSFR.

Access to Western components, technologies, and manufacturing techniques, sufficient investment, and direct entrepreneurial assistance are required to stimulate the development of a viable environmental products sector in Czechoslovakia.

INTRODUCTION

Like its neighbors Poland and Hungary, the Czech and Slovak Federal Republic (CSFR) is in the midst of transforming to market pluralism and a stable democracy after 42 years of Communist rule. Unlike Hungary, Czechoslovakia had no reform-minded communist party to ease the transformations. Unlike Poland, Czechoslovakia had no mass Solidarity trade union movement to challenge the old-line communist party's monopoly on power. Thus the CSFR, despite being one of the most prosperous pre-war democratic countries in Europe, is now "starting from scratch." In addition, the combination of the two nationalistic republics that comprise the Federal Union represents another significant difference between Czechoslovakia and it's neighbors to the north and south.

The long standing industrial tradition in Czechoslovakia resulted in the highest level of industrialization of any central and eastern European country in the pre-war era. The subsequent Communist command economy, emphasized heavy industry, mining, and building materials production.

The technologies used promoted extensive usage of energy and materials and produced high volumes of wastes. The command economy's emphasis on fulfillment of "the plan" encouraged industrial progress at the expense of resource efficiencies in energy and the environment. The Communist government of the CSFR paid lip service to the issue of environmental protection, while it's real priorities remained production and quotas. This resulted in a myriad of environmental transgressions and severe environmental degradation.

The Czech Republic, in the western half of the country, is one of Europe's biggest exporters of pollution. Dependence upon untreated low quality domestic coal for energy was, and still is, a major contributor to environmental degradation. During the Communist reign, the production of soft, brown, high-sulfur coal (lignite) increased approximately tenfold in the CSFR.

Czechoslovakia adopted legislation prior to 1980 that established the basis for setting ambient standards for most major pollutants, as well as monitoring, inspection and enforcement systems. Some of these standards were more rigorous than those found in western Europe. Unfortunately, the communist government largely ignored this legislation. However, recent federal environmental legislation indicates that the new leaders of the CSFR place significantly more emphasis on the environment than the previous regime.

Czechoslovakia is now developing an integrated strategy for addressing its myriad environmental problems. These problems include severe air pollution (sulfur dioxide, nitrous oxide, and particulates); water pollution caused by industrial and agricultural pollutants, and the discharge of untreated municipal wastewater; and an unknown number of hazardous and solid waste sites. This new strategy includes environmental legislation and regulations. It also addresses the domestic economic mechanisms required to fund the implementation of these environmental programs. However, the often divergent national and economic interests of the Czech and Slovak Republics complicates the process of developing and implementing a comprehensive environmental strategy in Czechoslovakia.

The Czech Republic contains 60% of the population and 70% of the industry. The Slovak Republic has a lower standard of living, and its industrial base focuses on production of military hardware and metallurgy. The growing nationalism in the two republics appears to threaten the very existence of the Federation. It seems likely that the ultimate price for keeping the country together will mean the acceptance of a much looser confederal structure. A looser confederal structure may impede establishing national environmental priorities and implementing a national environmental strategy for the Czech and Slovak Federal Republic.

The final component of a comprehensive environmental strategy necessary to protect and restore the environment of the CSFR is thriving domestic environmental business sectors. As one might expect, an economic system that gave protection of the environment a low priority, severely limited these business sectors prior to 1989. Today, as the CSFR moves away from a command toward a market economy, both the environmental goods and services sectors are developing and expanding in the two republics, albeit at greatly divergent rates.

New access to western technologies, technical assistance and training, and exposure to market and business management principles will be required to nurture and sustain the development of these two essential burgeoning business sectors.

The objectives of this report are to characterize the environmental market potential and the domestic capacity in the CSFR to satisfy those markets. It also identifies areas of opportunity for foreign assistance and investment to strengthen the domestic environmental goods and services sectors in Czechoslovakia.

ENVIRONMENTAL PRIORITIES IN CZECHOSLOVAKIA

Realistic estimates of the cost to protect and restore the environment of the CSFR over the next several decades are on the order of SU550-100 Billion. The national priorities identified below indicate that Czechoslovakia intends to control its pollution at the sources, through changes in resource efficiency and technology, rather than end-of pipe solutions.

AIR

- Reduction of sulfur dioxide and nitrogen oxide emissions;
- Coal cleaning and reduction of emissions from the combustion of soft coal;
- Reduction of air pollution from cars by introducing unleaded fuel, catalytic converters, and asbestos free brakes;

WATER

- Biological municipal wastewater treatment;
- Industrial and agricultural wastewater treatment;
- Installation of environmentally sound plumbing and piping to prevent loss of and accountability for potable water; and

WASTES

- Production of equipment for detection, collection, processing and storage of hazardous wastes;
- Segregation of municipal waste with recovery of raw materials; and
- Waste incineration and soil remediation.

As the CSFR implements its environmental programs, it will require a spectrum of environmental goods and services. Most of these products and services are not available in the domestic environmental sectors, especially in the Slovak republic. Over the next several years, the demand for the required goods and services will create sustainable environmental markets.

Today, Czechoslovakia faces two major challenges. First, it must fully develop those domestic skills, resources and assets that can lead to the development of competitive least cost environmental solutions. Second it must access the western technology, expertise and skills lacking in the CSFR in order to fulfill its long term environmental objectives.

DOMESTIC ENVIRONMENTAL MARKETS

STRUCTURE

Public concern about the environment and enaction and enforcement of environmental legislation creates demand for environmental goods and services. Czechoslovakia is evolving toward the European approach to environmental management involving a progression from law to standard-setting to negotiated permits to enforcement. However, the relevance of the conventional "western" market paradigm (FIGURE 1) to the analysis of the structure and potential of Eastern European environmental markets, like those in the CSFR, is dubious, because most components are absent.

The only paradigm component currently in place in every country in the region is the potential customer base, with its plethora of environmental needs. However, this component is not sufficient to drive dynamic environmental markets. The extent to which other necessary market mechanisms and forcing functions exist varies greatly between countries within the region. Each component of the market paradigm must be considered when analyzing and characterizing the environmental markets in each Eastern European country.

Eastern European countries are developing the institutional capability to promulgate, implement and enforce environmental regulations. In the CSFR, the Federal Committee for the Environment, the Ministry of the Environment of the Czech Republic, and the Slovak Commission for the Environment have only existed for 1-2 years. Each is organizing and building the institutional capabilities to carry out their respective missions. While the Federal and the Republican governments have agreed upon the division of responsibilities, agreement on the practical implementation of the responsibilities still gives rise to considerable dispute. Also, the delineation of environmental responsibilities between Republican and local governments has not yet been addressed.

Unlike Poland, the CSFR has not yet completed development of its comprehensive environmental strategy. As a result, the environmental legislation and economic funding mechanisms required to structure and fund environmental programs also are not yet clearly defined.

Finally, the political will to make environmental issues equal to economic issues associated with transformation is the most important variable in developing vital domestic environmental markets in the region. At the Federal level, political will appears to be gaining significant momentum in the CSFR and, among the countries in the region, is second only to that of Poland. However, the "competence" issues that exist among the three CSFR environmental government organizations will be difficult to resolve. These issues will likely work to impede development of the regulatory and economic mechanisms necessary to stimulate the environmental business sectors in Czechoslovakia.

The emerging domestic environmental markets in Czechoslovakia mirror the chaotic and rapid transition occurring throughout the economy and country. Characterizations of these markets may be obsolete before they are complete. Accurate estimates of the current environmental markets do not exist. However, the current environmental markets in the CSFR are certainly small relative to the current and future needs. Therefore, these markets are forecasted to grow at staggering (by U.S. standards) rates over the next decade.

ACTIVITY

Reliable estimates of the total level of environmental expenditures in the CSFR are difficult to identify. Perhaps the most comprehensive estimate can be derived from a 1990 survey (1) of "firms" (organizations) in the CSFR. The survey asked each organization to quantify the total turnover (revenues) derived from "environmental activities." The survey defines environmental in the broadest sense to include research and development, as well as consulting, processing of wastes, and environmentally sound energy production. The 1990 survey data *do not include* revenues derived from the production of environmental measurement technologies and other types of environmental products. The survey indicates that the total expenditures, in the CSFR, for

Figure 1
Schematic Diagram of the Conventional
Western Paradigm for Environmental Markets

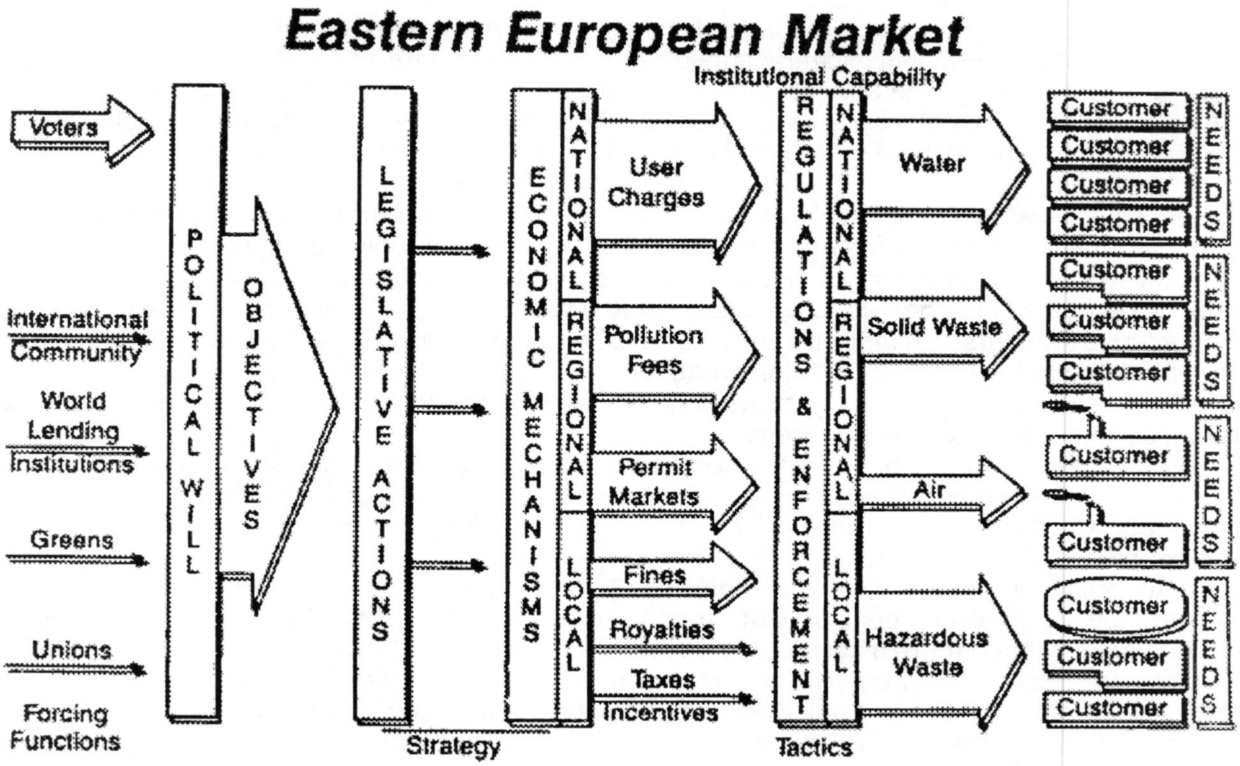

environmental services during 1990 was about $US 100 million. Approximately 70% of the market was in the Czech Republic and 30% in the Slovak Republic. The author estimates that the products sectors added another $US 35-45 million to the environmental market in Czechoslovakia.

Another measure of the current environmental markets in the CSFR considers the past and current government expenditures in these areas. For the period 1985-1990, the total reported environmental expenditures in the CSFR were approximately $US 500-600 million, an average of approximately $US 100-120 million pert year. For 1991, the Czech Republic budgeted approximately $US 200 million for environmental expenditures; the Slovak Republic had a comparable budget of $US 100 million. The amount of budgeted 1991 funds actually expended on environmental investments is not known. One can assume significant increases in environmental expenditures in the CSFR between 1990 and 1991. It is unlikely, however, that the environmental markets tripled, as the budgeted figures suggested.

The data suggest that, in the past, the government funded the bulk of expenditures in the environmental services and products sectors. This is consistent with a similar observation in Poland, where historically the government funded more than 80% of environmental expenditures.

The level of environmental expenditures in the CSFR over the next few years appears to depend on two major factors. The first is the level of economic activity, which may contract in the near term as restructuring occurs. The second is the level of pollution charges and fines, which may increase over the near term as increased emphasis on control and abatement develops. It is difficult to predict the net effect of these two forces in the near term. Clearly, long term prospects for the environmental business sectors in the CSFR are extremely positive.

FINANCING

A major concern regarding the future environmental products and services sectors in the CSFR is the government's ability to finance its goals. International organizations such as the World Bank, the European Bank for Reconstruction and Development, US A.I.D. and the EC have provided tens of millions of dollars in grants and loans to the CSFR for environmental programs.

In addition, government bilateral agreements, also contribute to these efforts. For example U.S. A.I.D. recently provided Czechoslovakia with a $US 12 Million sector grant to be used to implement environmental programs in the CSFR. The major objectives of such international aid are (a) institutional strengthening; (b) technology transfer; and (c) installation of pollution monitoring systems.

However, in view of the level of future environmental expenditures required in the CSFR, the role of international financial aid must be limited. Such aid can only serve as a catalyst to the domestic markets for environmental products and services. The major sources of funds must come from resources within the economy. Czechoslovakia intends to rely on basically two types of economic instruments to finance its environmental programs. The first instrument is pollution fees and fines aimed at promoting pollution abatement and discouraging new sources of pollution. The second instrument is user charges designed to both promote efficient use of environmental resources and reduce demand for them.

In the future, one might expect a maximum of 30% of domestic funding will come from the government. The remainder will have to come from the private sector. This will represent a major shift from the historical sources of environmental investments, but is consistent with the transformation to a market economy.

Czechoslovakia has adopted the *Polluter Pays Principle* (PPP). Under PPP, individual polluters are financially responsible both for the equipment to meet environmental requirements and for compensation for damages to the environment. There are, however, limitations to the PPP. Required enforcement is intensive, and it is difficult to monitor discharges and of polluters, or potential polluters. Also, some polluters do not have the financial resources required to fulfill their individual obligations under the law. Lastly, it is difficult to make the current polluters pay for the past neglect. This issue is a major concern in Eastern Europe, especially in the CSFR.

Equity and financial realism require public sources of funds to address certain environmental issues (e.g. abandoned waste sites). The Polluters Pay Principle (PsPP) addresses this problem and relieves the general taxpayer of the additional tax burden. Under PsPP, charges to current polluters provide funds to finance that portion of environmental restoration that cannot be adequately financed under the PPP. Usually, the government establishes Environmental Funds that receive and administer the expenditure of such fees, taxes and fines.

Czechoslovakia intends that the main source of finances for water use should be user charges. Such charges will be designed to ensure self-supporting regional water management systems. The structure for charges will be designed to stimulate water conservation. These rates will be controlled by the market not by the state. Both the Czech and Slovak Republics anticipate establishing air and water pollution charges, water intake charges, and charges for solid waste disposal. Rates for such charges have two objectives: (a) to generate revenues for environmental programs; and (b) to change polluter behavior.

A potential problem with the proposed system in the republics is that they rely on centralized environmental management. It is impossible to determine a rates for user charges that would be flat for an entire country and still allow for economically efficient solutions. In practice, a system of regionally differentiated rates, reflecting local conditions and resource costs, is preferable.

The government proposes introducing environmental taxes on hydrocarbon fuels and other products that have negative impacts on the environment by 1993. Other mechanisms such as soft loans, tax allowances, reduced import tariffs, etc. might be utilized where appropriate to accomplish the government's goals relating to environmental restoration and protection in the CSFR.

ENVIRONMENTAL BUSINESS SECTORS

STRUCTURE

The environmental services sector in the CSFR is in a state of dynamic change. A few years ago, only a handful of firms operated in the environmental consulting and services sector. Today, 125-150 environmental firms operate in the services sector. An equal, or slightly smaller, number of firms are active in the environmental products sector. Whether or not companies actually manufacture products, or simply supply western products and technologies, is not always clear.

Some firms result from entrepreneurial activity. Others result from the fragmentation of large state manufacturing, engineering and technical consulting firms that often occurs during privatization. Some firms in the services sector, have successfully competed in international markets (e.g. Stavebnie Geologie (Aquatest), GeoIndustria, HDP).

Today, the environmental services sector consists mostly of engineering and consulting firms. These focus on geological, geotechnical and hydrogeological consulting, and the design of air pollution control, sewage treatment, and industrial waste treatment systems.

The environmental products sector is currently limited in the level of technology sophistication. Former western export policies limited the Eastern European Bloc's access to advanced technological instrumentation and systems. Also, the CSFR lacked fundamental market demand for many essential products during the Communist regime. This suggests great potential demand for "high tech" environmental products in the Eastern European countries. However, these countries appear to prefer least cost solutions, at least during the initial phases of their environmental restoration programs

Historically, air filtration devices and small packaged sewage treatment plants dominated the environmental products sector in the CSFR. Recently, treatment technologies for industrial wastewater appeared in the marketplace. Lacking are many other types of equipment and devices (e.g. process control systems, field sampling and measuring devices, laboratory instrumentation, etc.)

Czechoslovakia is one of the few countries in Eastern Europe that appears to have a well organized and pro-active trade association for suppliers of environmental products and services. The Association of Producers of Environmental Systems (APES), located in Ostrava has over 100 members active in the environmental management, restoration and protection of all media.

ENVIRONMENTAL SERVICES

OVERVIEW

The technical base for an environmental services sector has existed in Czechoslovakia for many decades, but is now in a state of flux. The changes are due both to a spirit of entrepreneurism and the effects of the privatization process. Entrepreneurism leads to the creation of small specialized firms. More than one-third of the environmental services companies responding to the above referenced survey had less than 20 employees in 1990. Privatization may result in some service firms with more critical mass and a greater diversity of services, or larger specialized organizations.

However, Czechoslovakia requires significant investment in training, technology and time before its environmental services sector is comparable to, and competitive with, western counterparts. There are several areas of expertise associated with more contemporary approaches to environmental protection currently lacking in this sector.

For example, the emerging environmental strategy in the CSFR will emphasize the principle of sustainable development. It will encourage firms to rely on clean technologies, resource efficiencies, pollution prevention and waste minimization in designing their production process. It will discourage "end-of-the-pipe" control technologies. These innovative control approaches will create significant demand for specialized consulting and engineering expertise. Today, Czechoslovakia lacks the expertise to implement many of these innovative programs.

The technical strength of the CSFR in the environmental services sector may lie in it's world class hydrological, geological, hydrogeological and geotechnical expertise. In addition, Czechoslovakia has qualified engineers to design and construct water and wastewater treatment facilities, and, to a lesser extent, air filtration systems. This talent pool exists in state run specialized engineering organizations, focused on areas such as sanitary engineering, mining and metallurgy, wastewater and air particulates control, etc.

Also, Czechoslovakia's excellent academic and research institutions support a cadre of environmental scientists (ecologists, water quality experts, ecotoxicologists, geohydrologists, public health experts, etc.) Most of these scientists have mainly been involved in basic research, and may find the transition to applied science (i.e. consulting) difficult. Also (like their engineering counterparts), they lack the organizational, management, marketing and financial skills required to operate in a free market economy.

Many technical skills currently exist to develop a diverse environmental services sector in the CSFR. Access to organizational, business development, financial, marketing and management skills, specialized technical training (e.g. environmental auditing, waste minimization, pollution prevention, hazard and risk assessment), and key western technologies (remote sensing, soil and water remediation, etc.), should enable the CSFR to satisfy the environmental consulting services needs of its economy and society.

The geographical distribution of firms specializing in environmental services between the Czech and Slovak Republics is startling. Less than 20 of the approximately 140 firms identified in the environmental services sector are currently located in the Slovak Republic. This distribution ratio (i.e. 7:1) does not reflect the ratio of expenditures for environmental services between the two republics (i.e. 2:1). As might be expected, the major concentration of service firms is in Prague, with a secondary concentration in Brno.

There are very few large firms that offer general design, engineering, construction and/or consulting services within a broad sphere of environmental protection (e.g. air, water, wastewater and/or waste management). The exceptions are AB a.s. (Prague), HDP (Prague), GEOGMS s.p. (Prague), KMP-P (Prague), GFB (Brno) in the Czech Republic, and GEOKONZULT, s.p. (Kosice), in the Slovak Republic.

The focus of most firms is on hydraulic engineering, sanitary wastewater treatment, air pollution control, and/or waste management. Conspicuous by their absence are large numbers of firms advertising capability in industrial wastewater treatment, potable water treatment, pesticide and toxics monitoring, etc.

AIR POLLUTION

The largest firms in the area of air pollution control are ZVVZ (Milvesko) and AB a.s. (Prague). Both specialize in the design and installation of separators and air filtration plants. Smaller firms also specializing in this area are EKOENGINEERING (Brno), ILD (Prague), PBS BRNO, all located in the Czech Republic. No large firms were identified in the Slovak Republic. Only three Slovak firms were identified that are active in the area of air pollution. Each of these offered analytical services for air emissions measurement. The author assumes that such services are provided to Slovak industries by the suppliers in the Czech Republic.

There are a significant number of firms in the CSFR engaged in providing air analyses and monitoring capabilities. Approximately 35 firms were identified in the Czech republic, in addition to 3 in the Slovak Republic. The largest of these appear to be W AnCh (Usti nad Labem), VUHU (Most), and PUDIS (Prague) in the Czech Republic; and EKOLOGICKA in the Slovak Republic (Bratislava). There also some firms that specialize only in air quality measurements such as SZP (Tisice) and W SP (Prague) in the Czech Republic and UDI (Bratislava) in the Slovak Republic.

WATER MANAGEMENT & PROTECTION

The field of water management and pollution control has been, and is predicted to continue to be the largest segment of the environmental sector. Although exact figures for this market segment are not available, approximately 50% of environmental investments will be made in this area. Czechoslovakia's priorities in water protection include water, municipal sewage, and industrial wastewater treatment.

In the Czech Republic, 2500 municipalities, serving 2.5 million people have no domestic wastewater treatment. There are a approximately 30 firms offering basic consulting and engineering services in municipal sewage treatment (none in the Slovak Republic) and water purification. By far the largest of these are HDP (Prague), AB a.s. (Prague), SIGMA ENGINEERING s.p. (Olomouc), EKOINGSTAV BRNO s.p., INGSTAV HRADEC KRALOVE s.p., and VS-PIZ (Prague) in the Czech Republic; and STAVOPROJEKT (Kosice) in the Slovak Republic.

Several firms design and install small packaged sewage treatment plants for buildings and homes. These include AGROSTAV Sumperk a.s., AVOS Zamberk, and FITECH (Prague) in the Czech Republic and TEKOSPOL a.s. (Zilina) in the Slovak Republic. ELECTROSPOJ Prague) provides automatic control systems for municipal sewage treatment plants.

More than two dozen firms specialize in water supply and distribution. HYDROPROJECT, POVODI VLTAVY and AQUATEST (Prague), POVODY MORAVY and GEOTEST (Brno), POVODI ODRY (Ostrava), and VODNE ZDROJE and HYDROCONSULT (Bratislava) specialize in developing and protecting groundwater resources for supplying drinking water.

A few firms specialize in industrial wastewater treatment. The largest among these are VUK-PLZEN, VUTZ (Dvur Kralove Nad Labem) and K-SERVIS PRAHA s.p. in the Czech Republic, and W L Modra in the Slovak republic.

There are a number of firms which offer water and wastewater monitoring and analysis services. The largest of these are METRA Blansko, SPOLCHEMIE a.s. (Usti nad Labem), VUHU (Most), GEOINDUSTRIE CERNOSICE s.p., PUDIS (Prague), AQUATEST (Prague), and SCHB (Prague) in the Czech Republic, and VUKI (Bratislava) in the Slovak Republic.

WASTE MANAGEMENT

In 1987, Czechoslovakia produced more than 150 million tons of solid wastes, 80% were industrial wastes, and almost 20% are known to be hazardous wastes. Municipal solid waste represented only 3-4% of the solid wastes in the country.

Some studies of hazardous waste sites have been compiled in Czechoslovakia, but no comprehensive list or priority ranking scheme exists. Major problems include heavy metals from mine tailings, a variety organics from industry, and hydrocarbons, solvents, and other unknown contaminants from former Soviet Military installations.

No hazardous waste disposal sites are currently approved in Czechoslovakia. Identification, characterization and prioritization of sites, and risk based evaluations of human and ecological health are critical before cleanup can begin.

Many environmental service firms (almost half) are involved in waste management to one degree or another. Almost all of the previously mentioned generic consulting firms offer consulting services to refuse storage operations and for hazardous waste treatment. In addition to those firms, INGSTAV HRADEC KRALOVE s.p. and WPS (Prague) both located in the Czech Republic, and Edico (Bratislava) in the Slovak Republic also offer such services. EROS (Prague) provides liners and containment devices for waste sites. GFB (Brno) specializes in monitoring waste disposal sites.

Several firms provide consulting services in the recovery and recycling of various types of materials. KMP-P (Prague), INTES DRUSTVO PRO VYZKUM (Tabor-Mesice) and SUB (Brno) offer services in the recycling of building wastes. INORGA o.z. Ostrava offers services in the recycling of plastic wastes, while W K (Prague) and BIJO RECYCLING KONSORCIUM (Prague) offer services in the recovery of nonferric metals. ECOLOGIA (Prague) recycles petroleum wastes. No environmental services firms in the Slovak Republic appear to be involved in recovery and/or recycling.

VUAB (Melnik) designs and builds incinerators for both municipal and hazardous wastes.

Several firms have recently begun to specialize in the remediation of contaminated groundwater and soils. The leading firm in this sector is AQUATEST, (Prague). VSCHT PRAHA r.o. specializes in bioremediation of soil and groundwater.

OTHER SERVICES

A large number of firms (22) provide a broad spectrum of analytical laboratory services exclusively for analyzing environmental samples. For many of these firms, analytical services appears to be their principal source of revenues. Among these are AQUACONSULT (Prague), AGROSTAV (Prague), AGSS (Prague), GEOINDUSTRIA CERNOSICE s.p., CSUP-V W (Straz pod Ralskem), SZULP, (Ceske Budejovice), UNIGEO s.p. OSTRAVA, W HU (Most) W OS (Pardubice) and VARILAB (Prague) in the Czech Republic. The leading analytical services firms in the Slovak Republic are VUKI and UDI (both in Bratislava). A number of Czech firms provide very specialized measurement services. VUHU (Most), GEOINDUSTRIA CERNOSICE, GFB (Brno) and UDZ-CSAV (Prague) all provide radon and other radiometric analyses. SDPZ and GISAT (both located in Prague) provide remote sensing services, while the latter also provides Geographical Information Systems (GIS). W KL and VARILAB (both in Prague) offer pesticide and analyses of other contaminants in food samples, while EGO Chvaletice focuses on the analysis of fly ash.

KLUB ZELENEHO TELEFONU k.s.p. in Prague provides information and data bases on the environmental business sectors in Czechoslovakia. This firm developed the data base which forms the foundation of this report.

Appendix I presents a list of firms providing environmental services in Czechoslovakia, and indicates their major areas of consulting and/or engineering activity.

ENVIRONMENTAL PRODUCTS

OVERVIEW

Preliminary data indicate that the total market value of environmental products sold in the CSFR during 1991 was approximately $US 40 million. The breadth of the products sector appears extremely limited. Czechoslovakia does not manufacture many of the products and technologies that are critical to the implementation of its environmental restoration and pollution control strategies. For example, in the water treatment sector, there are few products produced in Czechoslovakia. In the air pollution control sector, there is a dearth of catalytic converters and technologies to abate pollution from the use of brown coal.

The sector with the largest number of domestic products is the wastewater treatment sector. This sector principally consists of packaged wastewater treatment plants (and their associated pumps, aerators, flocculants and coagulants). A reasonable number of firms also manufacture equipment for sampling, sensing and measuring chemical contaminants, but the level of technology and sophistication is probably not comparable to western products.

The same skewed distribution of suppliers that was evident in the services sector is apparently true for the products sector. Only about 15% of the suppliers of environmental product identified in Czechoslovakia are located in the Slovak Republic.

The environmental products sector may reflect the prior lack of markets for such products more than the capability (or lack thereof) of the manufacturing sector in the CSFR. There are several manufacturing sectors that appear to have the capabilities to provide many of the environmental products that will be required in the future. For example, the medical instrumentation sector appears relatively advanced and products apparently function adequately. In addition, there is emerging an electronics components and instruments (e.g. electronic testing) sector. Both these manufacturing sectors appear to have the potential to respond to these emerging environmental equipment markets.

The competitiveness of products manufactured in the CSFR suffers from both lower technology due to past isolation from the west, and higher costs due to management and resource inefficiencies. However, virtually every observer felt confident that, given access to technology, the technical expertise available in the manufacturing sector could produce the products required in the environmental sector.

There is concern about management expertise and ability to organize and motivate the "command economy" work force in existing organizations. Yet most observers were optimistic that the privatization process, by incentivizing the work force through ownership, would result in quality and productivity that eventually would compare favorably with western standards.

AIR POLLUTION

The priority areas for air pollution control center around reduction of sulfur dioxide emissions and reduction of dust. In addition, the government intends to expand its national monitoring network and is encouraging production of equipment for such programs.

CIPRESS FILTER (Brno), EKOENGINEERING (Brno) and ZVVZ (Milvesko) are the largest suppliers of equipment to this segment. SKODA PRAHA, ZVVZ (Milvesko), ZVU HRADEC KRALOVE s.p., OTES A. G. (Prague), and HUTNI MONTAZE OSTRAVA a.s. also provide equipment for flue gas desulphurization, systems for the control of nitrous oxides, and for dust control.

WATER TREATMENT

Few products appear to be manufactured domestically for this important sector. Poliz & Treoz (Brno) is the largest supplier of products to this sector, providing equipment for water filtration and purification. PLYNOSTAV (Prague) manufactures water pipelines. ZW HRADEC KRALOVE s.p. provides membranes for reverse osmosis systems for water treatment.

WASTEWATER TREATMENT

The government's environmental strategy includes massive investment in municipal and industrial sewage treatment to provide cities and industry with water of sufficient quality. In addition to packaged treatment plants, other equipment priorities include sludge dewatering and control equipment and systems for wastewater treatment plants.

The market for wastewater treatment and protection is already large (compared to other markets for environmental products). Though the domestic production capacity for plants and equipment is advanced, the low quality of equipment produced is a major problem. The principal suppliers of equipment to this market segment are AGROSTAV SUMPERK, KF (Havlickuv Brod) and CKD PRAHA, each providing a variety of aeration devices, pumps, filter presses, etc.

HUTNE STAVBY s.p. (Trencin) provides communal wastewater treatment systems designed to serve 1,000-10,000 inhabitants. W CHZ a. s. (Brno) provides small packaged aerobic sewage treatment plants to serve communities of 500-700 inhabitants. AIRCONSULT (Brno) specializes in the manufacture of fluidized bed systems for processing sludges, particulates etc. from the food processing industry. PREROVSKE STROJIRNY a.s. (Prerov) provides for filter presses for sludge dewatering. KF s.p. (Havlickuv Brod) provides microfiltration and ultrafiltration systems for industrial wastewater treatment. SIGMA ENGINEERING s.p. (Olomokic) provides a spectrum of pumps and pumping stations.

WASTE MANAGEMENT

Czechoslovakia's priorities in solid waste management focus on producing equipment for the collection, transportation, processing and storage of waste; segregation of waste with recovery of raw materials; composting; and construction of incineration plants. The principal domestic suppliers this segment are WASTEKO OSTROV nad OHRI, which is a principal producer of waste incineration systems, and CKD PRAHA, which provides a variety of mechanical systems for handling municipal wastes.

HUTNI MONTAZE OSTRAVA a.s. and USTAV NEROSTNYCH SUROVIN s.p. (Kutna Hora-Sedlec) provides waste recycling equipment and facilities. PREROVSKE STROJIRNY a.s. (Prerov) provides crushing and screening equipment for the recycling of waste building materials. AGROPROJECT PARDUBICE s.p. provides municipal waste sorting and recycling equipment and waste incinerators, and operates sanitary landfills.

CDK DUKLA (Prague) is a major supplier of municipal and industrial incinerators and boilers. PREROVSKE STROJIRNY a.s. (Prerov) produces world class high temperature rotary kilns to the cement industry in a western type facility. These kilns could easily applied to hazardous waste incineration. W SKLO UNION TEPLICE a.s. (Prague) provides vitrification systems for the treatment of incinerator ash. AIRCONSULT (Brno) provides fluidized bed systems for treating solid wastes from the food processing industry.

There are a reasonable number of suppliers of measurement equipment in the CSFR. The leading suppliers are DIGITEC (Prague), ELEKO PRAHA, CKD TECHNICKE (Prague) and ELINS (Slusovice), all located in the Czech Republic.

Appendix II presents a list of firms providing environmental products in Czechoslovakia, and indicates the major areas of product application and utility.

CONCLUSIONS

The last decade in Eastern Europe is testimony that serious environmental problems, per se, do not create environmental markets. A chain of events, involving the government, the people, and various segments of the private sector, must occur, on a national scale, before viable environmental markets emerge.

Capitalists like to think that efficient market forces are sufficient to internalize all true costs, even environmental costs. However, care for the environment is not a byproduct of market economies. Western industrialized societies have discovered that market forces, in and of themselves, are not sufficient to create environmental markets. On the contrary, every western industrialized society has had to rely on a hybrid regulatory approach consisting of command and control elements to establish objectives, and market mechanisms to provide least cost solutions. Czechoslovakia and the other emerging democracies of Eastern Europe must do the same.

In view of the myriad economic problems facing countries in transition from a command to market economy, a propensity to place environmental issues on the "back burner" is common. However, until political will to internalize environmental costs exists within a government, there is unlikely to be significant progress in addressing the environmental ills which plague countries in economic transition.

The government of the Czech and Slovak Federal Republic, as well as the legislative bodies in the two republics appear to have the political will necessary to address the countries environmental problems. Surely the national interests of the Republics, and the attendant issues related to competency, will create delays in implementing programs. However, despite these complicating issues, the cognizant governments have demonstrated the resolve to:

(a) enact enabling environmental legislation where required;
(b) develop a comprehensive environmental policy and strategy that integrates national, regional and local interests;
(c) identify and establish appropriate economic mechanisms to fund the country's environmental programs; and
(d) build the institutional capabilities required to develop and enforce regulations, and allocate resources.

In light of the legislative and institutional development occurring in Czechoslovakia, the development of a dynamic, domestic environmental products and services sectors seems inevitable.

The private sector in the CSFR is slowly emerging. There are some areas of expertise, and many essential environmental technologies, systems and products that are not yet available domestically. However, the potential to provide the required services exists, especially in the former state technical organizations and in Czechoslovakia's many fine academic institutions. The potential for domestic development of the required environmental products and technologies appears to be more limited. This sector will require significant foreign investment and technical support before it will be able to satisfy the domestic demand.

Organizational, business development, financial, marketing and management training, and some specialized technical skills training are essential to the development of a strong domestic environmental sector.

Some specialized technical training (e.g. waste minimization, pollution prevention, energy efficiency, risk assessment, etc.) are required to strengthen the domestic environmental services sector.

Access to western components, technologies, and manufacturing techniques, sufficient investment, and direct entrepreneurial assistance will stimulate the development of a viable products sector.

Appendix I
Domestic Environmental Services Suppliers

Company	Air	Water	Waste Water Treatment		Waste Treatment		
(Location)	Pollution	Supply	Industrial	Municipal	Industrial	Municipal	Other
AB a.s. xx (Praha 4)	xx		xx				
AB-geo Bratislava (Bratislava I)					xx		
ACHP Melnik (Melnik)							xx
AGP-C (Praha 1)						xx	
Agrostav Hronov (Nachod)					xx		
AGROSTAV spolecny podnik Litomerice (Litomerice)				xx			xx
AGROSTAV spolecny podnik Praha (Praha-vychod)		xx					
AGROSTAV Sumperk a.s. (Sumperk)				xx			
AGSS (Praha 2)							xx
AlphaBio International Ltd. (Praha 10)				xx			
AQUACONSULT (Praha-vychod)			xx				

Appendix I (*Continued*)
Domestic Environmental Services Suppliers

Company (Location)	Air Pollution	Water Supply	Waste Water Treatment Industrial	Waste Water Treatment Municipal	Waste Treatment Industrial	Waste Treatment Municipal	Other
AQUATEST (Praha 1)					xx		
AQUATIS s.p. (Brno-mesto)		xx		xx			xx
AVOS Zamberk (Usti nad Orlici)				xx			
B.O.I.S. spol. s r.o. (Brno-mesto)				xx			
Barumprojekt s.p. Zlin (Zlin)	xx						xx
BIJO Recycling, spol. s r.o. (Praha 10)					xx	xx	
BPT (Teplice)				xx			xx
CEI (Praha 4)							xx
Chemoprojekt Praha (Praha 2)				xx			xx
CHES (Usti nad Labem)							xx
CSAO (Hradec Kralove)			xx				
CSUP, PUUP-VTEI (Karlovy Vary)				xx	xx		
CSUP Uranove doly Dolni Rozinka odstepny zavod (Zdar nad Sazavou)	xx				xx		
CSUP-VVU (Ceske Lipa)	xx		xx	xx	xx		xx
ECOPLANT spol. s r.o. (Praha 5)	xx					xx	
EDICO (Bratislava I)	xx						xx
EGO Chvaletice (Pardubice)					xx		
EKO-service Praha, inzenyrska a zprostredkovatelska kancelar (Praha 2)				xx			

Appendix I (*Continued*)
Domestic Environmental Services Suppliers

Company (Location)	Air Pollution	Water Supply	Waste Water Treatment Industrial	Waste Water Treatment Municipal	Waste Treatment Industrial	Waste Treatment Municipal	Other
EKO-SERVIC Hradec Kralove (Hradec Kralove)	xx				xx	xx	xx
EKOARCH, atelier tvorby zivotniho prostredi (Plzen-sever)							xx
EKOBIOPROGRES Brno (Brno-mesto)			xx				
EKOENGINEERING (Brno-mesto)	xx						
EKOINGSTAV Brno s.p. (Brno-mesto)		xx		xx	xx		
EKOLOG (Brno-venkov)						xx	
EKOLOGIA (Praha 10)					xx		
EKOPLAN (Praha 6)							xx
EKOS (Praha 6)					xx		
EKOSPOL a.s. (Zilina)				xx			
Ekotechnik Praha (Praha 4)					xx		
Elektrospoj (Praha 8)				xx			
FESTING a.s. (Hradec Kralove)					xx	xx	xx
FITECH (Praha)			xx				
GEOAKTIV (Praha 4)					xx		
GEOGMS s.p. Praha (Praha 7)					xx		xx
Geoindustria Cernosice s.p. (Praha-zapad)					xx		xx
GEOKONZULT s.p. (Kosice-mesto)					xx		xx

Appendix I (*Continued*)
Domestic Environmental Services Suppliers

Company (Location)	Air Pollution	Water Supply	Waste Water Treatment Industrial	Waste Water Treatment Municipal	Waste Treatment Industrial	Waste Treatment Municipal	Other
GEOTEST (Brno-mesto)	xx						
GFB (Brno-mesto)	xx				xx		xx
GISAT (Praha 1)							xx
Grak (Praha 8)							xx
GT (Bratislava I)					xx		
HDP (Praha 4)	xx		xx	xx			
HYDROCONSULT (Bratislava I)		xx					
IC-04 (Praha 1)							xx
ILD xx (Praha 6)							
IMADOS (Praha 3)					xx		
INCOM s.p. (Praha 8)							xx
Ing. Milan Kriz, podnik pro zahranicne ochodni zastoupeni (Praha 4)							xx
Ingstav Hradec Kralove (Hradec Kralove)				xx	xx		
INORGA o.z. Ostrava (Ostrava-mesto)					xx		
INTES druzstvo pro vyzkum poradenske a inzenyrske sluzby (Tabor)					xx		
IP (Praha 1)							xx
K servis (Praha 5)		xx				xx	

Appendix I (*Continued*)
Domestic Environmental Services Suppliers

Company (Location)	Air Pollution	Water Supply	Waste Water Treatment Industrial	Waste Water Treatment Municipal	Waste Treatment Industrial	Waste Treatment Municipal	Other
KAP (Praha 6)	xx						
KF (Havlickuv Brod)		xx					
Klub Zeleneho telefonu, konsorcium (Praha 4)				xx			
KMP-P xx (Praha 6)			xx	xx	xx	xx	
KOMEKO a.s. (Kladno)		xx					
Kovoprojekta Praha s.p. (Praha 1)	xx	xx		xx			xx
KP Brno (Brno-mesto)						xx	
KPU (Praha 7)							xx
KUTEX EF (Blansko)	xx						
MAVEL a.s. (Praha 4)				xx			xx
METRA Blansko s.p. (Blansko)		xx					
OMaG (Jicin)							xx
ORGALEN s.p. (Trutnov)	xx						xx
P.M. Ing. Petr Mondschein (Praha 8)		xx		xx			
PBS Brnoxx (Brno-mesto)							
PEDOHYG Bratislava (Bratislava I)							xx
PGP (Praha 4)	xx		xx		xx		xx
PLASTSERVIS a.s. (Zlin)							xx

Appendix I (*Continued*)
Domestic Environmental Services Suppliers

Company	Air	Water	Waste Water Treatment		Waste Treatment		
(Location)	Pollution	Supply	Industrial	Municipal	Industrial	Municipal	Other
Poliz a Treoz (Brno-venkov)		xx					
PORCOS, spol. s r.o. (Bratislava-vidiek)							xx
POVOODIMORAVY (Brno-mesto)		xx					
POVODI ODRY (Ostrava-mesto)		xx					
POVODI VLATAVY (Praha 5)		xx					
Projekcni atelier ZA (Praha 8)							xx
Projektaxx (Praha 8)				xx	xx	xx	
PROSPECTA (Liberec)				xx			
PROVO (Svitavy)			xx				
PUDIS (Praha 8)	xx				xx		xx
REGULA Praha a.s. (Praha 4)	xx	xx		xx			
RVM (Most)							xx
SAEX>>a.s. (Praha 8)							xx
SCHB (Praha 5)					xx		
SDPZ (Praha 7)							xx
SEPARA (Brno-mesto)			xx				
SEVEROPROJEKT projektova a inzenyrska organizace (Ceske Lipa)					xx		
SG-PROEKO (Praha 1)					xx		

Appendix I (*Continued*)
Domestic Environmental Services Suppliers

Company (Location)	Air Pollution	Water Supply	Waste Water Treatment Industrial	Waste Water Treatment Municipal	Waste Treatment Industrial	Waste Treatment Municipal	Other
SIGMA ENGINEERING s.p. (Olomouc)				xx			
Sigma Hranice s.p. (Prerov)			xx				
SPES (Praha 6)							xx
SPOLCHEMIE a.s. (Usti nad Labem)	xx						
Statni rybarstvi Mestec Kralove s.p. (Nymburk)				xx			
Stavoprojekt (Kosice-mesto)				xx			
SUB (Brno-mesto)					xx		
SVB (Brno-mesto)	xx						
SZP (Melnik)	xx						
SZP (Praha 10)	xx			xx	xx	xx	
SXU Brnoxx (Brno-mesto)			xx				
SZULP (Ceske Budjovice)	xx	xx					
TEKOSPOL a.s. (Zilina)				xx			
UACH CSAV (Praha 6)	xx						xx
UDI (Bratislava I)	xx						xx
UDI Praha (Praha 1)							xx
UDZ CSAV (Praha 8)							xx
UEL-SAV (Zvolen)						xx	

Appendix I (*Continued*)
Domestic Environmental Services Suppliers

Company (Location)	Air Pollution	Water Supply	Waste Water Treatment Industrial	Waste Water Treatment Municipal	Waste Treatment Industrial	Waste Treatment Municipal	Other
UG CSAV (Praha 8)		xx					
UNBR CSAV (Praha 4)							xx
UNIGEO s.p. Ostrava (Bruntal)	xx						
UNISOF v.o.s. (Jablonec nad Nisou)				xx	xx		xx
URBANPLAN x.r.o. (Hradec Kralove)							xx
URS Praha (Praha 2)				xx			xx
USMD (Praha 1)	xx						xx
Ustav vyuziti plynu Brno (Brno-mesto)					xx	xx	xx
UTIA CAV (Praha 8)							xx
UTZCHT CSAV (Praha 6)					xx		
UUG (Praha 1)					xx		
UVTIZ (Praha 2)							xx
VARILAB spol. s r.o. (Praha 1)	xx						xx
VHS (Zilina)				xx			
VODNE ZDROJE (Praha 1)		xx					
VS-PIZ (Praha 7)	xx		xx	xx			
VSCHT PRAHA r.o. (Praha 1)					xx		
VSUOZ (Praha-zapad)							xx

Appendix I (*Continued*)
Domestic Environmental Services Suppliers

Company (Location)	Air Pollution	Water Supply	Waste Water Treatment		Waste Treatment		Other
			Industrial	Municipal	Industrial	Municipal	
VTX (Praha 4)							xx
VUAB (Melnik)					xx	xx	
VUAnChxx (Usti nad Labem)				xx			
VUGPT s.p. (Zlin)							xx
VUHU (Most)	xx						xx
VUK (Praha-vychod)					xx		
VUK Plzen (Plzen-mesto)			xx				
VUKI (Bratislava I)	xx					xx	xx
VUKL (Praha 4)	xx						xx
VUL Modra (Bratislava-vidiek)			xx				xx
VUMO (Praha 5)	xx						
VUMOP (Praha 5)				xx		xx	
VUOS (Pardubice)	xx						xx
VUPS (Praha 10)					xx		xx
VUPU (Bratislava I)							xx
VUSP (Praha 1)	xx						
VUTZ (Trutnov)			xx				
VUV T.G.M. Praha (Praha 6)		xx	xx				

Appendix I (*Continued*)
Domestic Environmental Services Suppliers

Company (Location)	Air Pollution	Water Supply	Waste Water Treatment Industrial	Waste Water Treatment Municipal	Waste Treatment Industrial	Waste Treatment Municipal	Other
VUZORT (Praha 5)						xx	
Z-projekt (Praha 4)						xx	
Zahradni architektura s.p. Kvitkuv Dvur (Cesky Krumlov)							xx
ZD "Zalesi" (Zlin)							xx
Zemedelska druzstvo Dzbanov (Praha 10)							xx
ZETOR s.p. (Brno-mesto)			xx				
ZS C. Budejovice (Ceske Budejovice)				xx	xx		
ZUNZ UP-UHP UP (Pribram)							xx
ZVVZ (Pisek)	xx						

Appendix II
Domestic Environmental Products Suppliers

Company (Location)	Air Pollution	Water Supply	Waste Water Treatment Industrial	Waste Water Treatment Municipal	Waste Treatment Industrial	Waste Treatment Municipal	Other
AGROPROJECT PARDUBICE (Pardubice)					x x		
Agrosluzby, druzstevni podnik (Zdar nad Sazavou)						x x	
AGROSTAV Sumperk a.s. (Sumperk)		x x	x x	x x			
AGS Brno-venkov (Brno-venkov)			x x				
AIRCONSULT (Brno-mesto)			x x		x x		
AO (Trutnov)			x x				

Appendix II (*Continued*)
Domestic Environmental Products Suppliers

Company (Location)	Air Pollution	Water Supply	Waste Water Treatment Industrial	Waste Water Treatment Municipal	Waste Treatment Industrial	Waste Treatment Municipal	Other
Aquatech (Beroun)				x x			
Bio-Stav a.s. (Praha 4)							x x
BRIKLIS konzorcium (Tabor)					x x	x x	
BSR DELTA AG Basilej, zastoupeni Praha (Praha 1)		x x					
CHZS (Sokolov)			x x	x x			
CIPRES FILTR (Brno-mesto)	x x						
CKD E, a.s. (Praha 9)	x x						
CKD Praha a.s. (Praha 9)		x x	x x	x x	x x	x x	
CKD TL a.s. (Praha 9)	x x						
CSAO (Hradec Kralove)			x x				
CSUP Uranove doly Dolni Rozinka, odstepny zavod (Zdar nad Savavou)			x x				
DIGITEC, konstruckcni kancelar (Pribram)		x x	x x	x x			
EGO Chvaletice (Paradubice)	x x						
EKO (Nachod)	x x						
EKO-SERVIS Hradec Kralove (Hradec-Kralove)	x x						x x
EKOBIOPROGRES Brno (Brno-mesto)							x x
EKOENGINEERING (Brno-mesto)	x x						

Appendix II (*Continued*)
Domestic Environmental Products Suppliers

Company (Location)	Air Pollution	Water Supply	Waste Water Treatment Industrial	Waste Water Treatment Municipal	Waste Treatment Industrial	Waste Treatment Municipal	Other
ELEKO (Praha 9)	x x						x x
ELINS (Zlin)		x x	x x	x x			x x
GEOspol s.p. (Brno-mesto)		x x	x x	x x			
GFB (Brno-mesto)							x x
GT (Bratislava I)					x x	x x	
HS Kosice (Kosice-mesto)				x x			
HURT-vyroba ohrivacu (Prachatice)							x x
HUTNE STAVBY s.p. (Trencin)				x x			
HUTNI MONTAZE OSTRAVA (Ostrava-mesto)	x x				x x	x x	
Ing. Milan Kriz, podnik pro zahranicne ochodni zastoupeni (Praha 4)		x x	x x	x x			x x
JANKA s.p. (Praha 5)	x x						
JET SERVICE (Praha 6)			x x				
KF (Havlickuv Brod)			x x		x x	x x	
KOTA Trencin (Trencin)	x x						
KOVOPODNIK Pardubice s.p. (Pardubice)		x x	x x	x x			
LPP (Praha 6)							x x
MAVEL a.s. (Praha 4)							x x
MITOP a.s. (Ceske Lipa)	x x				x x	x x	

Appendix II (*Continued*)
Domestic Environmental Products Suppliers

Company (Location)	Air Pollution	Water Supply	Waste Water Treatment Industrial	Waste Water Treatment Municipal	Waste Treatment Industrial	Waste Treatment Municipal	Other
OIKOS-Burian (Plzen-mesto)			x x				
OTES A. G. (Praha 1)	x x						
PAL-Magneton s.p. (Kromeriz)		x x	x x	x x			
PBS Brno (Brno-mesto)	x x						x x
PLYNOSTAV (Praha 1)		x x					
POLA (Praha 2)					x x	x x	
Poliz a Treoz (Brno-venkov)		x x					
PREROVSKE STROJIRNY (Prerov)			x x		x x		
PROSPECTA (Liberec)		x x					
SENSIT, v.o.k. (Vsetin)							x x
SEPARA (Brno-mesto)		x x	x x	x x			
SG-PROEKO (Praha 1)					x x	x x	
SIGMA ENGINEERING s.p. (Olomouc)				x x			
SKODA PRAHA (Praha 1)	x x						
Statni statek Most (Most)							x x
STS Hodonin (Hodonin)			x x			x x	
SVUT (Liberec)		x x					
TANEX a.s. (Nachod)		x x					

Appendix II (*Continued*)
Domestic Environmental Products Suppliers

Company (Location)	Air Pollution	Water Supply	Waste Water Treatment Industrial	Waste Water Treatment Municipal	Waste Treatment Industrial	Waste Treatment Municipal	Other
TMS s.p. Pardubice (Paradubice)			x x	x x	x x		
UNEX (Olomouc)					x x	x x	
UNSTAV NEROSTNYCH SUROVIN (Kutna Hora)					x x	x x	
UTZCHT CSAV (Praha 6)	x x						
VaS EKOSYSTEM vyrobni projektova organizace (Praha 9)			x x				
VS-SZ (Tabor)			x x				
VUSKLO UNION TEPLICE (Praha 1)					x x		
VUAnCh (Usti nad Labem)	x x						
UVCHZ a.s. (Brno-mesto)				x x			
VUHU (Most)	x x						
VUTZ (Trutnov)		x x	x x	x x			
VVULP (Plzen-mesto)			x x				
VZUCHOTECHNIKA a.s. (Trencin)	x x						
WASTEKO s.p. ekologicke strojirenstvi (Karlovy Vary)	x x				x x	x x	
ZD-CHOTANKY (Nymburk)		x x	x x	x x			
ZD "Javornik" (Novy Jicin)		x x	x x	x x			
ZD Mir Potehy (Kutna Hora)	x x						

Appendix II (*Continued*)
Domestic Environmental Products Suppliers

Company (Location)	Air Pollution	Water Supply	Waste Water Treatment Industrial	Waste Water Treatment Municipal	Waste Treatment Industrial	Waste Treatment Municipal	Other
ZD Podlesi (Vsetin)					x x		
Zemedelske druzstvo v Kobyli na Morave (Breclav)	x x			x x	x x		
ZTS s.p. (Lucenec)						x x	
ZVU HADREC KRALOVE (Hradec Kralove)	x x	x x					
ZVVZ (Pisek)	x x				x x		

APPENDIX J - MARKETING PLAN: POLAND

This appendix was made available by the U.S. and Foreign Commercial Service, Warsaw, Poland. It is dated November 1991.

I. COUNTRY DATA

A. Profile and Contacts

Population: 39 Million
Religion: 90% Roman Catholic
Government: Parliamentary Democracy
Language: Polish
Work Week: 42 Hours

Key Contacts in Warsaw (Telephone: 48/22/21-45-15, Fax: 21-63-27)
Ms. Joan J. Edwards, Senior Commercial Officer
Mr. Dale N. Tasharski, Commercial Officer

Key Contacts in Washington
Ms. Audrey Zuck, Poland Desk Officer (Telephone: 202-377-4915)
Mr. Charles Ford, Regional Director (Telephone: 202-377-1599)
Ms. Susan Blackman, Director, East European Business Information Center (Telephone: 202-377-2645)

B. Poland's Domestic Economy (USDOL Millions, Except as Noted)

Category	1989	1990	1991(E)
GDP Growth Rate (% of Prev. Yr.)	.2%	-12%	-10%
GDP Per Capita (in USDOL)	2710	1675	1500
Gov't Spending (% of GNP)	N/A	4%	6%
Inflation (% of Prev. Yr.) (+)	330%	638%	70%
Unemployment (Percent)	N/A	3.4%	11%
Foreign Exchange Reserves	2500	4900	4650
Average Exchange Rate (*)	1150	9500	11200
Foreign Debt	40000	40850	44300
Capital Debt Reduction	0	0	365
Eco-Conversion Debt Reduction	0	0	160
U.S. Econ. Assistance (@)	0	N/A	202
U.S. Military Assist. (#)	0	0	.5
U.S. PL-480 Spendings (&)	N/A	10	4
U.S. Charitable Donations	0	0	150

Notes:
\+ Central Planning Office of Poland
E Estimated, Unless Otherwise Noted
N/A Data Not Available
* Polish Zloty Per USDOL
@ Obligated Value, A.I.D., through August 1991
# International Military Education & Training (IMET)
& National Bank of Poland

C. Trade Data for Poland (USDOL Millions, Except as Noted)

Category	1989	1990	1991(E)
Total Exports	13469	15503	1300
Total Imports	10279	9230	12500
Exports to the U.S.	387	371	350
Imports from the U.S.	414	406	300
U.S. Share of Polish Imports (%)	4%	4%	2%
Imports of Manufactured Goods			
From All Countries	8651	6629	6400
From the U.S.	136	325	250
U.S. Share, Manf. Imp. (%)	2%	5%	4%
Trade Balances, Rank of Leading Partners in 1990			
1. Germany	+37	+1471	+968(*)
2. United Kingdom	+52	+421	+429(*)
3. USSR	+118	+382	-366(*)

Leading U.S. Exports to Poland in 1990

Aircraft and Equipment ($79 Million)
Maize, Unmilled ($38 Million)
Oil, from Petroleum ($15 Million)
Wheat, Unmilled ($14 Million)
Animal Feeds ($10 Million)

Leading U.S. Imports from Poland in 1990

Meat ($82 Million)
Mens and Boys Coats and Jackets ($23 Million)
Tractors ($20 Million)
Fresh Fish, Chilled or Frozen ($17 Million)
Copper ($16 Million)

Notes:
(Data Calculated at Current 11,500 Zloty per USDOL Rate)
E Estimated, Unless Otherwise Noted
N/A Data Not Available
* Actual Balance Through August 1991

D. Investment in Poland (USDOL Millions, Except as Noted)

Category	1989	1990	1991(E)
Total Foreign Direct Investment	N/A	215	650
U.S. Direct Investment	N/A	15	80
U.S. Share of Foreign Investment	N/A	7%	12%

Foreign Country Investments, Rank by Number of Investments (*)

	Country, Number	Value	% of Total
1.	Germany, 1446	$153 Million	23%
2.	United States, 394	53 Million	8%
3.	Sweden, 340	46 Million	7%
4.	Austria, 311	36 Million	5%
5.	Holland (@), 274	45 Million	7%
6.	France, 245	65 Million	10%
7.	United Kingdom, 241	35 Million	5%
8.	Italy, 223	33 Million	5%

Major U.S. Investors in Poland (Alphabetical Order)

AT&T
Boeing
Gerber
IBM
PepsiCo Foods International

———————

Notes:
(Data Calculated at Current 11,500 Zloty per USDOL Rate)
* Through 2nd Quarter
N/A Data Not Available
@ Benelux Countries

Sources for All Country Data:
Department of Commerce Statistical Data
Poland Statistical Yearbook, 1990
Polish Agency for Foreign Investment
Statistical Bulletin, Poland, October 1991, No. 10
Foreign Trade Bulletin, Poland, August 1991

II. BEST PROSPECTS

A. General Comments

This section is comprised of those industry sectors that represent what USFCS Warsaw believes are the top 20 "Best Prospects" (most promising sectors) for U.S. exports of technology and service to Poland in 1992. We rank each best prospect by its potential for U.S. business in Poland, based upon objective analyses of the market conditions and available statistical data.

It is important to note that sectorial statistical data are not readily available in Poland. Data are drawn from numerous sources and contacts, and should not be considered consistent (for comparative analysis) across all subsectors, except where U.S. statistical data are provided (i.e. U.S. export data). Further, all data are estimates. Growth rate figures are subject to significant change, given the rapid changes in Poland's developing market economy. A comprehensive source of Poland's developing market economy. A comprehensive source of Polish statistical sectorial data is simply not yet available. Most promising subsectors are indicated in each best prospect, based upon consultation with industry sources. Poland has implemented the Harmonized System and this should lead to consistent subsector data in the future.

Poland is in its economic infancy. Here, past is not prologue. Markets may seem small, but each has potential for significant long-term growth. USFCS Warsaw urges those companies interested in doing business in the following best prospect sectors to examine each prospect in terms of the overall economic and business conditions in Poland.

B. List of Best Prospects

Each best prospect is named and coded for commercial reporting purposes. The size of the Polish market for products in each sector, the value of all imports of products in the sector, and the value of U.S. exports to Poland in the sector are estimated for the calendar year 1991. Estimated growth figures are also provided. Each best prospect includes a numerical ranking for Polish receptivity, domestic and foreign competition, and trade obstacles for that sector. Comments are provided, and a list of obstacles for that sector. Comments are provided, and a list of most promising subsectors for each best prospect is included.

Importantly, customs duty rates and turnover tax rates (applied to the import value of the product *plus* the customs duty) are provided for each best prospect in the comments section. Customs rates are subject to change and suspension.

A. Rank of Sector: 1
B. Name of Sector: Telecommunications Equipment
C. Code: TCE
D. Est. 1991 Market Size (USDOL Millions): 550
E. Est. Average Annual Growth Rate of Market (Percent): 50%
F. Est. 1991 Total Imports (USDOL Millions): 150
G. Est. Average Annual Growth Rate of Total Imports (Percent): 75%
H. Est. 1991 Imports from the U.S. (USDOL Millions): 15
I. Est. Average Annual Growth Rate of Imports from the U.S. (Percent): 300%
J. 1. Poland's Receptivity to U.S. Products in this Industry (1=Lowest; 5=Highest) 5
   2. Competition for U.S. Exporters in this Sector from Polish and Foreign Suppliers (1=Heavy; 5=Very Little) 3
   3. Overall Effect of Polish Governmental and Private Sector Trade Barriers on U.S. Exports in this Sector (1=Very Severe; 5=Very Few) 5

K. Comments:
(8517) Poland is expected to invest over $12 billion in its telecommunications industry through the year 2000; current total credits allocated equal $816 million. This is perhaps the hottest best prospect for U.S. technology in Poland. The sector is entirely underdeveloped, and the entire country, from big city to small, is looking to modernize. There is fierce competition for major bids, and U.S. technology must be adaptable to European standards. Plans for the next two years include 12 intercontinental automatic transit exchanges, 16 radio links, and 14 fiber optic links. The market for telephone apparatus and small telephone exchanges is currently showing sudden growth.
   1. Customs Duty: 15%
   2. Turnover Tax: -0- (25% for telephones)

L. Most Promising Subsectors (Est. 1991 Market Size in USDOL Millions):
   1. Parts of Telephone Lines (360)
   2. Unspecified Telephone Apparatus (70)
   3. Carrier-Line Systems (40)
   4. Telephones (35)
   5. Switching Equipment (35)

A.	Rank of Sector:	2
B.	Name of Sector:	Food Processing & Packaging Equipment
C.	Code:	FPP
D.	Est. 1991 Market Size (USDOL Millions):	420
E.	Est. Average Annual Growth Rate of Market (Percent):	15%
F.	Est. 1991 Total Imports (USDOL Millions):	300
G.	Est. Average Annual Growth Rate of Total Imports (Percent):	6%
H.	Est. 1991 Imports from the U.S. (USDOL Millions):	1
I.	Est. Average Annual Growth Rate of Imports from the U.S. (Percent):	60%
J.	1. Poland's Receptivity to U.S. Products in this Industry (1=Lowest; 5=Highest)	5
	2. Competition for U.S. Exporters in this Sector from Polish and Foreign Suppliers (1=Heavy; 5=Very Little)	3
	3. Overall Effect of Polish Governmental and Private Sector Trade Barriers on U.S. Exports in this Sector (1=Very Severe; 5=Very Few)	5

K. Comments:
(8422, 8509) New customs regulations are expected to slow the growth of imports of food products into Poland over the next few years. Poland will look to modernize its current food production facilities and open new product lines. U.S. exporters of heavy industrial food processing machinery will realize their potential as Poland's state-owned food factories continue to privatize,and Western investment pours in.
   1. Customs Duty: 15-20%
   2. Turnover Tax: 0-25%

L. Most Promising Subsectors (Est. 1991 Market Size in USDOL Millions):
   1. Confectionery Machinery (130)
   2. Packaging Equipment (100)
   3. Bottling Equipment (70)
   4. Fruit and Vegetable Processing (35)

A. Rank of Sector: 3
B. Name of Sector: Management and Consulting Services
C. Code: MCS
D. Est. 1991 Market Size (USDOL Millions): 50
E. Est. Average Annual Growth Rate of Market (Percent): 50%
F. Est. 1991 Total Imports (USDOL Millions): 35
G. Est. Average Annual Growth Rate of Total Imports (Percent): 75%
H. Est. 1991 Imports from the U.S. (USDOL Millions): 10
I. Est. Average Annual Growth Rate of Imports from the U.S. (Percent): 60%
J. 1. Poland's Receptivity to U.S. Products in this Industry (1=Lowest; 5=Highest) 5
   2. Competition for U.S. Exporters in this Sector from Polish and Foreign Suppliers (1=Heavy; 5=Very Little) 2
   3. Overall Effect of Polish Governmental and Private Sector Trade Barriers on U.S. Exports in this Sector (1=Very Severe; 5=Very Few) 5
K. Comments:
The U.S. government has placed highest emphasis on programs for management and training in Poland. Poland began a free market economy, based on Western business practice, from scratch in 1990. As Poland privatizes state-owned industry, the demand for management and consulting services is proving great. The Polish government recognizes the need to restructure and modernize as quickly as possible, in order to trade and compete with the world, and with that recognizes the need for training. Newly privatized firms are entering into business transactions only to find the need for advice and counsel. The largest U.S. accounting firms are represented in Poland, as are a growing number of larger U.S. law firms.
L. Most Promising Subsectors (Est. 1991 Market Size in USDOL Millions):
   1. Banking Services (20)
   2. Accounting Services (15)
   3. Marketing Services (2)

A. Rank of Sector: 4
B. Name of Sector: Printing and Graphic Arts Equipment
C. Code: PGA
D. Est. 1991 Market Size (USDOL Millions): 20
E. Est. Average Annual Growth Rate of Market (Percent): 35%
F. Est. 1991 Total Imports (USDOL Millions): 10
G. Est. Average Annual Growth Rate of Total Imports (Percent): 35%
H. Est. 1991 Imports from the U.S. (USDOL Millions): 1
I. Est. Average Annual Growth Rate of Imports from the U.S. (Percent): 50%
J. 1. Poland's Receptivity to U.S. Products in this Industry (1=Lowest; 5=Highest) 5
   2. Competition for U.S. Exporters in this Sector from Polish and Foreign Suppliers (1=Heavy; 5=Very Little) 2
   3. Overall Effect of Polish Governmental and Private Sector Trade Barriers on U.S. Exports in this Sector (1=Very Severe; 5=Very Few) 5
K. Comments:
(8442, 8443) Customs duty is applied to printing and graphic arts equipment in Poland, but not printed materials. Therefore, many Polish firms have printed abroad. The democratization of Poland has witnessed a surge in publishing activity, and new newspapers and journals seem to appear daily, some in foreign languages. However, a lack of quality paper and modern technique is apparent, and there is room for growth. Small U.S. printing machinery companies

have had some early success in 1991.
1. Customs Duty: 15%
2. Turnover Tax: - 0 -

L. Most Promising Subsectors (Est. 1991 Market Size in USDOL Millions):
1. Typesetting Equipment (8)
2. Parts for Print Blocks (4)
3. Letterpress Printing Machinery (3)
4. Other Printing Machinery (3)

A. Rank of Sector: 5
B. Name of Sector: Pollution Control Equipment
C. Code: POL
D. Est. 1991 Market Size (USDOL Millions): 20
E. Est. Average Annual Growth Rate of Market (Percent): 5%
F. Est. 1991 Total Imports (USDOL Millions): 15
G. Est. Average Annual Growth Rate of Total Imports (Percent): 5%
H. Est. 1991 Imports from the U.S. (USDOL Millions): 2
I. Est. Average Annual Growth Rate of Imports from the U.S. (Percent): 5%
J. 1. Poland's Receptivity to U.S. Products in this Industry (1=Lowest; 5=Highest) 4
2. Competition for U.S. Exporters in this Sector from Polish and Foreign Suppliers (1=Heavy; 5=Very Little) 3
3. Overall Effect of Polish Governmental and Private Sector Trade Barriers on U.S. Exports in this Sector (1=Very Severe; 5=Very Few) 5

K. Comments:
(8421) U.S. pollution control equipment, through respected for its technological capabilities, does not yet have the presence nor acceptance in Poland than European manufacturers have. In cost competitive situations, the preference generally leans to those firms that provide the most support for their products. There is a perception among end-users in this sector that U.S. firms are too distant geographically to provide fast and comprehensive service to their equipment. Overcoming this barrier will result in greater receptivity and more sales for U.S. pollution control equipment manufacturers.
1. Customs Duty: 15%
2. Turnover Tax: - 0 -

L. Most Promising Subsectors (Est. 1991 Market Size in USDOL Millions):
1. Filtering/Purifying Equipment for Water (9)
2. Filtering/Purifying Equipment for Other Liquids (5)
3. Filtering/Purifying Equipment for Gases (5)

A. Rank of Sector: 6
B. Name of Sector: Air Conditioning & Refrigeration Equipment
C. Code: ACR
D. Est. 1991 Market Size (USDOL Millions): 80
E. Est. Average Annual Growth Rate of Market (Percent): 15%
F. Est. 1991 Total Imports (USDOL Millions): 30
G. Est. Average Annual Growth Rate of Total Imports (Percent): 4%
H. Est. 1991 Imports from the U.S. (USDOL Millions): 1
I. Est. Average Annual Growth Rate of Imports from the U.S. (Percent): 50%
J. 1. Poland's Receptivity to U.S. Products in this Industry (1=Lowest; 5=Highest) 5

2. Competition for U.S. Exporters in this Sector from Polish and Foreign Suppliers (1=Heavy; 5=Very Little) 4
3. Overall Effect of Polish Governmental and Private Sector Trade Barriers on U.S. Exports in this Sector (1=Very Severe; 5=Very Few) 3

K. Comments:
(8415, 8418, 8419) U.S. manufacturers of air conditioning, refrigeration, and heating equipment enjoy respect in Poland due to the quality of their products. U.S. products have excellent potential for growth in this sector, as we have seen significant growth in inquiries for air conditioning equipment over the past months. Wall units became popular this past summer, but are still expensive for the average Pole. U.S. exports of refrigeration sector over the last 18 months. Poland is modernizing its warehouse and distribution system to keep up with the heavy influx of imported food and consumer products. This should support growth in this sector for many months to come.
1. Customs Duty: 15-20%
2. Turnover Tax: 25%

L. Most Promising Subsectors (Est. 1991 Market Size in USDOL Millions):
1. Refrigerator Freezer Parts (65)
2. Heating Equipment (6)
3. Air Conditioners, Wall Units, and Parts (3)
4. Freezers (2)

A. Rank of Sector: 6
B. Name of Sector: Household Consumer Goods
C. Code: HCG
D. Est. 1991 Market Size (USDOL Millions): 850
E. Est. Average Annual Growth Rate of Market (Percent): 10%
F. Est. 1991 Total Imports (USDOL Millions): 55
G. Est. Average Annual Growth Rate of Total Imports (Percent): 25%
H. Est. 1991 Imports from the U.S. (USDOL Millions): 2
I. Est. Average Annual Growth Rate of Imports from the U.S. (Percent): 10%
J. 1. Poland's Receptivity to U.S. Products in this Industry (1=Lowest; 5=Highest) 5
2. Competition for U.S. Exporters in this Sector from Polish and Foreign Suppliers (1=Heavy; 5=Very Little) 2
3. Overall Effect of Polish Governmental and Private Sector Trade Barriers on U.S. Exports in this Sector (1=Very Severe; 5=Very Few) 5

K. Comments:
(3924, 7013, 8211, 8509, 8516, 9105, 9701) Poles spend a considerable amount of their income on household articles. Free market pricing in the household consumer goods sector made the price gap between foreign and domestic products much smaller, and opened the market for foreign made household electric equipment. German and French items like toasters and mixers have sold well in Poland. German and Italian plastic kitchenware are commonly available in the network of retail shops in the market. There is considerable room for continued import growth, and tremendous potential for U.S. products which can compete in quality and price.
1. Customs Duty: 15%
2. Turnover Tax: 20-25%

L. Most Promising Subsectors (Est. 1991 Market Size in USDOL Millions):
1. Household Plastic Articles (600)
2. Domestic Appliances (140)
3. Tableware of Glass, Ceramics and Metal (30)

A. Rank of Sector: 8
B. Name of Sector: Computers and Peripherals
C. Code: CPT
D. Est. 1991 Market Size (USDOL Millions): 250
E. Est. Average Annual Growth Rate of Market (Percent): 20%
F. Est. 1991 Total Imports (USDOL Millions): 230
G. Est. Average Annual Growth Rate of Total Imports (Percent): 20%
H. Est. 1991 Imports from the U.S. (USDOL Millions): 20
I. Est. Average Annual Growth Rate of Imports from the U.S. (Percent): 20%
J. 1. Poland's Receptivity to U.S. Products in this Industry (1=Lowest; 5=Highest) 5
   2. Competition for U.S. Exporters in this Sector from Polish and Foreign Suppliers (1=Heavy; 5=Very Little) 2
   3. Overall Effect of Polish Governmental and Private Sector Trade Barriers on U.S. Exports in this Sector (1=Very Severe; 5=Very Few) 4
K. Comments:
(8471) Computer hardware and software are excellent prospects for U.S. exports to Poland. The hardware market is dominated by IBM clones of Far Eastern origin and is extremely price competitive for U.S. products, while the microcomputer market is dominated by European subsidiaries of major manufacturers. 90% of current procurements are currently being made by state enterprises and factories. The Polish government has been and is still a good prospect for sales. There are six computer related trade fairs each year in Poland, and a variety of international computer journals available.
   1. Customs Duty: 20%
   2. Turnover Tax: - 0 -
L. Most Promising Subsectors (Est. 1991 Market Size in USDOL Millions):
   1. Computers (100)
   2. Printers (90)
   3. Hard Disc Drives and Tape Drivers (50)

A. Rank of Sector: 9
B. Name of Sector: Electrical Power Systems
C. Code: EPS
D. Est. 1991 Market Size (USDOL Millions): 25
E. Est. Average Annual Growth Rate of Market (Percent): 2%
F. Est. 1991 Total Imports (USDOL Millions): 5
G. Est. Average Annual Growth Rate of Total Imports (Percent): 5%
H. Est. 1991 Imports from the U.S. (USDOL Millions): 1
I. Est. Average Annual Growth Rate of Imports from the U.S. (Percent): 10%
J. 1. Poland's Receptivity to U.S. Products in this Industry (1=Lowest; 5=Highest) 4
   2. Competition for U.S. Exporters in this Sector from Polish and Foreign Suppliers (1=Heavy; 5=Very Little) 3
   3. Overall Effect of Polish Governmental and Private Sector Trade Barriers on U.S. Exports in this Sector (1=Very Severe; 5=Very Few) 4
K. Comments:
(8416, 8504, 8511, 9030) The energy sector is in desperate need for new, environmentally safe technologies. In addition, power facilities seek energy efficient equipment and systems. The United States is seen as a leader in both of these areas, and the potential of U.S. exports of this technology is significant. New legislation governing fines levied on polluters including power

plants, makes the need for modernization immediate.
1. Customs Duty: 15%
2. Turnover Tax: 25%

L. Most Promising Subsectors (Est. 1991 Market Size in USDOL Millions):
1. Voltage Current Regulation Equipment (20)
2. Turbines (2)

A. Rank of Sector: 10
B. Name of Sector: Hotel and Restaurant Equipment
C. Code: HTL
D. Est. 1991 Market Size (USDOL Millions): 20
E. Est. Average Annual Growth Rate of Market (Percent): 20%
F. Est. 1991 Total Imports (USDOL Millions): 10
G. Est. Average Annual Growth Rate of Total Imports (Percent): 40%
H. Est. 1991 Imports from the U.S. (USDOL Millions): .2
I. Est. Average Annual Growth Rate of Imports from the U.S. (Percent): 200%
J. 1. Poland's Receptivity to U.S. Products in this Industry (1=Lowest; 5=Highest) 5
2. Competition for U.S. Exporters in this Sector from Polish and Foreign Suppliers (1=Heavy; 5=Very Little) 2
3. Overall Effect of Polish Governmental and Private Sector Trade Barriers on U.S. Exports in this Sector (1=Very Severe; 5=Very Few) 4

K. Comments:
(8418, 8419, 8428, 8531, 9504) Though the statistics currently appear small for U.S. export sales, the potential for U.S. export sales in this sector is unlimited. Trade fair contacts for U.S. firms have been very high in volume. New restaurants are opening daily in major cities across Poland. Growth potential is already strong for U.S. exporters of small business food processing machinery, as entrepreneurialship is booming in Poland's restaurant sector, even during recession.
1. Customs Duty: 20%
2. Turnover Tax: 25%

L. Most Promising Subsectors (Est. 1991 Market Size in USDOL Millions):
1. Elevators (10)
2. Cold Storage Displays (3)
3. Burglar and Fire Alarms (3)
4. Cooking Ovens (2)

A. Rank of Sector: 11
B. Name of Sector: Furniture
C. Code: FUR
D. Est. 1991 Market Size (USDOL Millions): 300
E. Est. Average Annual Growth Rate of Market (Percent): 8%
F. Est. 1991 Total Imports (USDOL Millions): 40
G. Est. Average Annual Growth Rate of Total Imports (Percent): 30%
H. Est. 1991 Imports from the U.S. (USDOL Millions): 1
I. Est. Average Annual Growth Rate of Imports from the U.S. (Percent): 100%
J. 1. Poland's Receptivity to U.S. Products in this Industry (1=Lowest; 5=Highest) 5
2. Competition for U.S. Exporters in this Sector from Polish and Foreign Suppliers (1=Heavy; 5=Very Little) 2

3. Overall Effect of Polish Governmental and Private Sector Trade Barriers on U.S. Exports in this Sector (1=Very Severe; 5=Very Few) 4

K. Comments:
(9402, 9402) There is a tremendous potential for U.S. office furniture and equipment in Poland. Poland experienced a boom in imports of office furniture in late 1990 and early 1991, and this trend is continuing, as well as in shop/store displays and equipment. Poland's retail market is blooming in the face of recession, and newly opened and privatized businesses seek the latest in Western style furniture and equipment. The whole furniture industry in Poland is undergoing privatization (currently 95% state-owned), which will lead to increased opportunities for new types and styles, instead of the current traditional and Scandinavian designs. Cane and bamboo were exported to Poland from the U.S. last year.
   1. Customs Duty: 20%
   2. Turnover Tax: 25%

L. Most Promising Subsectors (Est. 1991 Market Size in USDOL Millions):
   1. Wooden Furniture, Not Office (140)
   2. Wooden Office Furniture (6)
   3. Metal Furniture (50)

A. Rank of Sector: 12
B. Name of Sector: Sporting Goods
C. Code: SPT
D. Est. 1991 Market Size (USDOL Millions): 100
E. Est. Average Annual Growth Rate of Market (Percent): 25%
F. Est. 1991 Total Imports (USDOL Millions): 45
G. Est. Average Annual Growth Rate of Total Imports (Percent): 35%
H. Est. 1991 Imports from the U.S. (USDOL Millions): 1
I. Est. Average Annual Growth Rate of Imports from the U.S. (Percent): 10%
J. 1. Poland's Receptivity to U.S. Products in this Industry (1=Lowest; 5=Highest) 4
   2. Competition for U.S. Exporters in this Sector from Polish and Foreign Suppliers (1=Heavy; 5=Very Little) 2
   3. Overall Effect of Polish Governmental and Private Sector Trade Barriers on U.S. Exports in this Sector (1=Very Severe; 5=Very Few) 5

K. Comments:
(9506, 9507) This is a potentially unlimited market for U.S. exports in the 1990's. Personal expenditures for sporting goods are increasing, and domestic production and imports are also on the rise. Equipment is imported from Germany and Australia, and several West European brands have established dealers in Poland. Until new sports are introduced in Poland, sales will continue in water recreation equipment, hunting, riding, and skiing equipment.
   1. Customs Duty: 20%
   2. Turnover Tax: 25%

L. Most Promising Subsectors (Est. 1991 Market Size in USDOL Millions):
   1. Fishing Equipment (40)
   2. Tennis Equipment and Balls (30)
   3. Gymnasium Equipment (15)

A. Rank of Sector: 13
B. Name of Sector: Medical Equipment
C. Code: MED
D. Est. 1991 Market Size (USDOL Millions): 50

E. Est. Average Annual Growth Rate of Market (Percent): 12%
F. Est. 1991 Total Imports (USDOL Millions): 25
G. Est. Average Annual Growth Rate of Total Imports (Percent): 4%
H. Est. 1991 Imports from the U.S. (USDOL Millions): 6
I. Est. Average Annual Growth Rate of Imports from the U.S. (Percent): 2%
J. 1. Poland's Receptivity to U.S. Products in this Industry (1=Lowest; 5=Highest) 4
   2. Competition for U.S. Exporters in this Sector from Polish and Foreign Suppliers (1=Heavy; 5=Very Little) 2
   3. Overall Effect of Polish Governmental and Private Sector Trade Barriers on U.S. Exports in this Sector (1=Very Severe; 5=Very Few) 4
K. Comments:
(9018) There is no functioning distribution system for medical equipment and apparatus in Poland. Hospitals, doctors, and sometimes even patients purchase needed equipment independently through a variety of direct sources. Plans for privatizing the health sector (which would open up avenues for distribution) are pending, as Poland waits for the passage of a law on public insurance. Private clinics and dental facilities are beginning to appear. Offering pay-for-services, these purchase new equipment based upon the needs of specific patients.
   1. Customs Duty: 15%
   2. Turnover Tax: - 0 -
L. Most Promising Subsectors (Est. 1991 Market Size in USDOL Millions):
   1. Diagnostic Equipment (20)
   2. Surgical and Medical Apparatus (20)
   3. Dental Equipment (5)

A. Rank of Sector: 14
B. Name of Sector: Industrial Process Controls
C. Code: PCI
D. Est. 1991 Market Size (USDOL Millions): 180
E. Est. Average Annual Growth Rate of Market (Percent): 2%
F. Est. 1991 Total Imports (USDOL Millions): 100
G. Est. Average Annual Growth Rate of Total Imports (Percent): 10%
H. Est. 1991 Imports from the U.S. (USDOL Millions): 1
I. Est. Average Annual Growth Rate of Imports from the U.S. (Percent): 50%
J. 1. Poland's Receptivity to U.S. Products in this Industry (1=Lowest; 5=Highest) 5
   2. Competition for U.S. Exporters in this Sector from Polish and Foreign Suppliers (1=Heavy; 5=Very Little) 2
   3. Overall Effect of Polish Governmental and Private Sector Trade Barriers on U.S. Exports in this Sector (1=Very Severe; 5=Very Few) 3
K. Comments:
(9032) Polish production in this sector has been decreasing, and imports are on the rise. Industry realizes the need for modernization and the technological benefits of process controls, but factories are currently strapped for capital. Imports are expected to show significant growth as industry privatizes and money becomes available. Companies supplying equipment through joint ventures have greater potential than those seeking import distribution at this time.
   1. Customs Duty: 15%
   2. Turnover Tax: 25%
L. Most Promising Subsectors (Est. 1991 Market Size in USDOL Millions):

1. Automatic Regulators (140)
2. Thermostats (18)
3. Hydraulic and Pneumatic Controls (12)

A. Rank of Sector: 15
B. Name of Sector: Consumer Electronics
C. Code: CEL
D. Est. 1991 Market Size (USDOL Millions): 930
E. Est. Average Annual Growth Rate of Market (Percent): 10%
F. Est. 1991 Total Imports (USDOL Millions): 310
G. Est. Average Annual Growth Rate of Total Imports (Percent): 70%
H. Est. 1991 Imports from the U.S. (USDOL Millions): 6
I. Est. Average Annual Growth Rate of Imports from the U.S. (Percent): 10%
J. 1. Poland's Receptivity to U.S. Products in this Industry (1=Lowest; 5=Highest) 4
   2. Competition for U.S. Exporters in this Sector from Polish and Foreign Suppliers (1=Heavy; 5=Very Little) 2
   3. Overall Effect of Polish Governmental and Private Sector Trade Barriers on U.S. Exports in this Sector (1=Very Severe; 5=Very Few) 5
K. Comments:
(8518, 8519, 8520, 8521, 8522, 8527, 8528) This will soon be a billion dollar market in Poland. There is great consumer demand for electronics in Poland. Domestic production has decreased in 1991, and imports have filled the void. The market is dominated by Japanese and relatively cheap Korean and Taiwanese products (color TV sets, VCRs, satellite receivers and antennas), and is extremely price sensitive on the low-end. Microwave ovens, CB radios, and telephone answering machines are new-to-market and selling very well. German products are very visible. The former retail monopoly of Pewex and Baltona stores no longer exists, as smaller retailers began to emerge in 1990.
   1. Customs Duty: 15-30% (some rates partially suspended)
   2. Turnover Tax: 25-30%
L. Most Promising Subsectors (Est. 1991 Market Size in USDOL Millions):
   1. TV Sets (450)
   2. Video Equipment and Parts (230)
   3. Tape Recording Equipment (110)

A. Rank of Sector: 16
B. Name of Sector: Travel and Tourism Services
C. Code: TRA
D. Est. 1991 Market Size (USDOL Millions): 450
E. Est. Average Annual Growth Rate of Market (Percent): 10%
F. Est. 1991 Total Imports (USDOL Millions): 20
G. Est. Average Annual Growth Rate of Total Imports (Percent): 10%
H. Est. 1991 Imports from the U.S. (USDOL Millions): .3
I. Est. Average Annual Growth Rate of Imports from the U.S. (Percent): 10%
J. 1. Poland's Receptivity to U.S. Products in this Industry (1=Lowest; 5=Highest) 4
   2. Competition for U.S. Exporters in this Sector from Polish and Foreign Suppliers (1=Heavy; 5=Very Little) 2
   3. Overall Effect of Polish Governmental and Private Sector Trade Barriers on U.S. Exports in this Sector (1=Very Severe; 5=Very Few) 3

K. Comments:
The tourism service market is expected to grow steadily through the 1990's. Construction of hotels, and sale of hotel space will be the early focus. Poland has great potential for growth in tourism services, with a base of outstanding but relatively underdeveloped resort areas. The Mazurian and Pomeranian lake districts, the Tatra mountain region, and the seaside of the north offer unique and unspoiled locations. Almost 95% of Poland's income in the tourist sector is earned by the state travel agency Orbis, which is currently being prepared for privatization. Small travel agencies offering tour packages are beginning to surface in the major cities.

L. Most Promising Subsectors (Est. 1991 Market Size in USDOL Millions):
1. Hotel Rooms and Services (50)

A. Rank of Sector: 17
B. Name of Sector: Automobile Parts and Service Equipment
C. Code: APS
D. Est. 1991 Market Size (USDOL Millions): 800
E. Est. Average Annual Growth Rate of Market (Percent): 5%
F. Est. 1991 Total Imports (USDOL Millions): 90
G. Est. Average Annual Growth Rate of Total Imports (Percent): 15%
H. Est. 1991 Imports from the U.S. (USDOL Millions): 10
I. Est. Average Annual Growth Rate of Imports from the U.S. (Percent): 20%
J. 1. Poland's Receptivity to U.S. Products in this Industry (1=Lowest; 5=Highest) 4
2. Competition for U.S. Exporters in this Sector from Polish and Foreign Suppliers (1=Heavy; 5=Very Little) 2
3. Overall Effect of Polish Governmental and Private Sector Trade Barriers on U.S. Exports in this Sector (1=Very Severe; 5=Very Few) 3

K. Comments:
(8708) Though competition is heavy in this sector, U.S. exports to Poland continue to flourish. Poland is grappling with its import policy for automobile parts, realizing it must keep duty rates low to attract foreign investment. There has been a tremendous turnout at local car trade fairs featuring cars parts and accessories. The automotive industry is expected to see continued strong growth in the years ahead, with new dealerships sprouting up across the country and new factories (with Western investment) being built in the major cities.
1. Customs Duty: 15%
2. Turnover Tax: 25% (0% if parts are used for assembly)

L. Most Promising Subsectors (Est. 1991 Market Size in USDOL Millions):
1. Unspecified Automobile Parts (700)
2. Radiators (30)
3. Gears and Boxes (20)

A. Rank of Sector: 18
B. Name of Sector: Construction Equipment
C. Code: CON
D. Est. 1991 Market Size (USDOL Millions): 120
E. Est. Average Annual Growth Rate of Market (Percent): 5%
F. Est. 1991 Total Imports (USDOL Millions): 20
G. Est. Average Annual Growth Rate of Total Imports (Percent): 25%
H. Est. 1991 Imports from the U.S. (USDOL Millions): 3
I. Est. Average Annual Growth Rate of Imports from the U.S. (Percent): 50%

J. 1. Poland's Receptivity to U.S. Products in this Industry (1=Lowest; 5=Highest) 3
2. Competition for U.S. Exporters in this Sector from Polish and Foreign Suppliers (1=Heavy; 5=Very Little) 1
3. Overall Effect of Polish Governmental and Private Sector Trade Barriers on U.S. Exports in this Sector (1=Very Severe; 5=Very Few) 5

K. Comments:
(8479, 8704, 8705) U.S. exports in this sector are greatly affected by the general recession in Poland and price sensitivities in the housing and construction market—a market that is as a whole underfinanced. On the surface, however, construction seems to be progressing. Opportunities for U.S. sales are expected to appear where modern equipment and methods are needed, but U.S. exporters will need to adjust their prices accordingly to increase market share.
1. Customs Duty: 15%
2. Turnover Tax: 0-25%

L. Most Promising Subsectors (Est. 1991 Market Size in USDOL Millions):
1. Unspecified Mechanical Machinery (55)
2. Parts for Mechanical Machinery (50)
3. Dump Trucks (7)

A. Rank of Sector: 19
B. Name of Sector: Textile Products: Finished Goods
C. Code: TXP
D. Est. 1991 Market Size (USDOL Millions): 1340
E. Est. Average Annual Growth Rate of Market (Percent): 20%
F. Est. 1991 Total Imports (USDOL Millions): 140
G. Est. Average Annual Growth Rate of Total Imports (Percent): 30%
H. Est. 1991 Imports from the U.S. (USDOL Millions): 5
I. Est. Average Annual Growth Rate of Imports from the U.S. (Percent): 10%
J. 1. Poland's Receptivity to U.S. Products in this Industry (1=Lowest; 5=Highest) 4
2. Competition for U.S. Exporters in this Sector from Polish and Foreign Suppliers (1=Heavy; 5=Very Little) 1
3. Overall Effect of Polish Governmental and Private Sector Trade Barriers on U.S. Exports in this Sector (1=Very Severe; 5=Very Few) 4

K. Comments:
(HS Chapters 61, 62) This sector has seen considerable expansion over the last 18 months. In 1990, exports were greater than imports, but this seems to have reversed in 1991. The majority of imports in this sector are from China, Korea, Hong Kong, and India, while Western European products are available too. U.S. products can compete in quality, but must be sensitive to price. U.S. garments are also sought after, but still relatively unavailable in Poland.
1. Customs Duty: 30%
2. Turnover Tax: 25%

L. Most Promising Subsectors (Est. 1991 Market Size in USDOL Millions):
1. Womens and Girls Clothing, and Underwear (1000)
2. Mens and Boys Clothing (200)

A. Rank of Sector: 20
B. Name of Sector: Automobiles and Light Trucks/Vans
C. Code: AUT
D. Est. 1991 Market Size (USDOL Millions): 2260

E. Est. Average Annual Growth Rate of Market (Percent): 20%
F. Est. 1991 Total Imports (USDOL Millions): 1220
G. Est. Average Annual Growth Rate of Total Imports (Percent): 20%
H. Est. 1991 Imports from the U.S. (USDOL Millions): 20
I. Est. Average Annual Growth Rate of Imports from the U.S. (Percent): 20%
J. 1. Poland's Receptivity to U.S. Products in this Industry (1=Lowest; 5=Highest) 5
   2. Competition for U.S. Exporters in this Sector from Polish and Foreign Suppliers (1=Heavy; 5=Very Little) 2
   3. Overall Effect of Polish Governmental and Private Sector Trade Barriers on U.S. Exports in this Sector (1=Very Severe; 5=Very Few) 2
K. Comments:
   (8703) In 1991 Poland will import a record number of cars, while domestic manufacture will continue to decline as Soviet and East European markets dry up for Polish automobiles. Some estimated indicated that up to $3 billion in imports are possible, with popular models such as Mercedes, BMW, Volkswagon, and Ford leading this growth. Used car imports are also showing growth, and will continue to do so in 1992. Used cars, however, must be converted to European specifications for import into Poland. Statistical estimates may be conservative.
   1. Customs Duty: 25%
   2. Turnover Tax: 20-30%
L. Most Promising Subsectors (Est. 1991 Market Size in USDOL Millions):
   1. Passenger Cars, Spark Ignited (1600)
      Greater than 1500 cc but less than or equal to 3000 cc)

III. COMMERCIAL ENVIRONMENT

Recent Polish history is dramatic and heroic. It reflects favorably on Poland. Efforts to remold the country into a Western market economy are on track, although fraught with unavoidable complications. There are no tested formulas for undoing the damage of 44 years of communist mismanagement.

Poland is attempting to cope with a shortage of competent civil servants, as it rushes headlong into building a market economy. 1990-1991 saw the election of Lech Walesa as President and the installation of a capable and energetic government under the direction of Prime Minister Jan Krzysztof Bielecki. A chief theme of the government has been the speeding up of legislative development of private businesses. The election of a new parliament, in late October 1991, and the subsequent adoption of a new constitution should complete the codification of a democratic system.

One of the most welcoming aspects of the new Polish market for U.S. exporters is that the government and people of Poland hold the United States in extreme high regard. American products enjoy a special image with both industrial and household consumers. The Ministry of Foreign Economic Relations notes that more U.S. firms (than those from any other country) have arrived to explore future business in Poland in the period since the new government took over in late 1989. Government ministers have also states that, given complete equality of technical merit and price, an American offer will routinely beat a European or Japanese proposal. Nevertheless, American exporters face sharp competition from German, French, British, Italian, Swedish, Austrian, and Far East manufacturers.

A. Economic Considerations

Poland made a dramatic return to parliamentary democracy in late 1989. Efforts to rebuild the economy in 1990 on the basis of free enterprise began in earnest. Economic change and political and social restructuring have continued at a staggering pace. Great optimism and great uncertainty, born of rapid change, are the predominant characteristics of the commercial environment.

1991 proved a particularly difficult year for the Polish economy, as rapid price adjustments cancelled out former subsidies, and other restructuring measures continued. There were major fluctuations in Poland's regional and international markets. Its trade with other former socialist partners and the Soviet Union moved to a hard currency basis. Under the direction of Deputy Prime Minister Balcerowicz, Poland earned respect in the global business arena for diligent implementation of an IMF economic stabilization program.

Unfortunately, the pains of economic austerity are being severely felt by average working class families, the very group which lies at the heart of President Walesa's broad popular support. The months ahead will witness tensions between the government and the working class—unemployment is expected to increase. Poland's government can be expected to continue to strive to meet IMF targets, but under increasing pressures from labor organizations.

Opportunities for foreign investment, hailed as one of the major engines of economic transformation in Poland, have drawn the interest of American and Western European business. Over 2000 joint ventures with foreign participation now exist. Following the passage of privatization legislation in July of 1990, and the establishment of a new ministry to oversee the process, Poland began a broad scale transformation of state-owned and municipal enterprises to shift the ownership of these companies to the private sector.

To protect their investments, U.S. investors now have available the programs of the Overseas Private Investment Corporation (OPIC), which began operation in Poland in late 1989. OPIC can provide to U.S. investors investment insurance against certain political risks and financing through direct loans, loan guarantee and equity investment.

B. Political and Legislative Factors

There are many developments which will affect the business environment in Poland over the next two years whose results cannot be predicted. The Solidarity movement, long held together by opposition to the old government, has split into distinct political factions. Fully free parliamentary elections take place in late October 1991. The elections will result in the formation of Poland's third post communist government in as many years. The issues facing the new government are how to continue serious economic reform and revitilization while continue addressing social concerns—health care, unemployment, housing.

Legislation needed to establish and regulate Poland's growing private economy is steadily being put in place. New laws on employment, wages, taxes, insurance, health care, foreign investment, customs, banking, securities and exchange, privatization, ownership, foreign exchange, telecommunications, mining, and intellectual property rights have passed or are expected to pass soon. Moreover, many new laws are subject to amendment or overhaul following their early introduction. Though adjusting to new rules can cause temporary business disruptions, the trend is clearly toward considerable improvement in the legislative framework over the medium and long term. For example, in June 1991, the Polish parliament passed a revised law on foreign investment which permits full repatriation of profits. At the same time it retains a liberal tax holiday option for large-scale investments:

1) in regions of high unemployment, or
2) in new technologies of national interest, or
3) where at least 20% of production is for exports

One important area undergoing rapid evolution is the regulatory framework for the privatization of large state enterprises. The law on state-owned privatization passed the parliament in late 1990.

Several early privatizations have been successfully completed. Foreign takeovers of Polish state enterprises are both legal and welcome. Complications often arise; none are completed quickly. U.S. firms interested in participating in the privatization of state industry in Poland should contact the Ministry of Privatization, assemble their own internal team of experts to study potential projects, and then complement the team with good Polish legal and technical consultants.

C. Commercial Factors

There is tremendous unrealized demand for Western goods in Poland which creates ample opportunity for large-scale exports by large, medium, and small American producers. While lack of cash and/or financing remain major limiting factors, there have been some surprising successes for U.S. exporters in the aircraft, medical equipment, food processing, telecommunications, computer, and air traffic control industries.

Massive foreign aid programs and unprecedented debt relief measures were put in place in early 1991. These greatly enhance the prospects for financial support for major projects identified by the Polish Government as national economic priorities.

The re-establishment of Eximbank credits, and the involvement of the World Bank in Poland's infrastructural development, creates significant opportunity for American exporters to participate in major projects in Poland over the next few years. Liberalization of Western export licensing rules for Poland opens new opportunities to exporters of U.S. high technology products.

Medium-sized U.S. exporters with quality products at competitive prices enjoy growing opportunities for sales in Poland. Poland has witnessed an explosion of retail small business and entrepreneurial growth, based largely on imported products following the legalization of the former "second economy" in early 1989. Small business development, a government priority and a cause supported by the Polish-American Enterprise Fund, has resulted in a proliferation of small firms seeking trade relationships with smaller Western businesses. Demand for services such as business consulting, accounting, and legal counsel present strong opportunities for U.S. firms.

There are a number of practical concerns when doing business in Poland. These include a notoriously underdeveloped telephone and postal system, an archaic banking system, a shortage of corporate lawyers, and a lack of up-to-date information on Polish firms currently in business. In the interim, U.S. business partners must adjust their style of operation to account for such deficiencies. Necessary restructuring and modernization is taking place in all of these areas and progress will be felt month-to-month over the next few years. U.S. companies seeking to establish offices in Poland will face a growing shortage in (and rising costs for) office space and experienced staff in 1992.

There are certain start-up costs for exporters and other businesses in Poland. Virtually all success stories begin with a patient and committed exploratory effort followed by a well thought out business proposal. Poland is an excellent market for experienced exporters ready to expand to new markets. Firms entirely new to exporting should regard Poland with some caution, and they should pursue trade relationships strictly on practical business realities.

Changes within the overall trading structure of Poland have produced a unique environment which can be difficult for new-to-market exporters. When the old state trading system was legally abolished in 1989, the 70 major trading firms lost their monopoly rights. They now face competition among themselves and from private and foreign firms. But they are experienced and remain dominant because they have old and well-established internal networks and contacts in overseas markets. As a result, it may still make sense for U.S. firms to establish contacts within these giant trading houses. At the same time, these firms continue to undergo significant organizational and functional changes. They should no longer be regarded merely as leftover elements of the communist system which created them.

Overall, USFCS Warsaw is optimistic about the long-term prospects for U.S. trade and investment in Poland. Both the risks and the potential profits are above average vis-a-vis other prospective overseas markets. Firms willing to get involved now will benefit from the substantial business relationship that is developing between the United States and Poland.

The U.S. Department of Commerce has programs and services, available through its District Offices, designed to help a U.S. exporter do business in Poland—and around the world. To get the telephone number of your local USFCS District Office, call the Commerce Department in Washington at (202) 377-4767. Additional business information on Poland can be obtained from the Eastern European Business Information Center (EEBIC) at the Commerce Department by calling (202) 377-2645.

IV. FINANCIAL ENVIRONMENT

The Polish government is serious about eliminating the practice of countertrade. The government recognizes that countertrade weakens market mechanisms and adds inflationary costs. Strict licensing requirements for countertrade transactions have been introduced. Polish firms that engage in countertrade must provide details on the volume and value of a transaction, commodity specifications, and a justification of why monetary payment cannot be made. These requirements are designed to make countertrade unattractive for trading firms. Licensing waivers apply in cased where surplus Polish products are traded for domestically scarce commodities or where the profitability of the transaction for the Polish partner is clear.

Poland has been severely burdened with heavy debt since the late 1970s. In March 1991, Western governments started a process to forgive 50 percent of Poland's official debt—providing Poland stays on track with its IMF economic stabilization program. This debt forgiveness (by the Paris Club of official lenders) has not been matched by the London Club of private creditors, although negotiations towards debt relief continue with major international banks.

Debt relief will make additional Polish funds available for investment. But investment funds from outside sources will continue to be limited and, normally, will be those that are integrated into assistance programs.

Private U.S. banks have remained reluctant to extend commercial credits or to guarantee Polish bank payments. Most financing is found under special programs such as those of the World Bank, the new European Bank for Reconstruction and Development (EBRD), and other financial assistance programs created by Western governments.

A. Methods of Payment for Exports

Import financing procedures adhere to Western business practice. All payments go through one of the following qualified foreign exchange banks whose guarantees are reliable: Bank Handlowy w Warszawie S.A., Bank Polska Kasa Opieki S.A. (Pekao, S.A.), Bank Rozwoju Eksportu S.A. (Export Development Bank), Powszechna Kasa Oszczednosci-Bank Panstwowy (PKO-BP), and Bank Gospodarki Zywnosciowej (Bank for Food Economy).

The safest method of receiving payment for a U.S. export is through an irrevocable letter of credit (L/C). Banks in Poland require the importer to deposit funds prior to issuance of a letter of credit. Most difficulties in obtaining U.S. bank guarantees on Polish letters of credit seem to stem from consideration of Poland's overall debt performance. They do not reflect the actual performance of Polish banks on L/C's, and there are no known cases of default on payment over a period of the last five years. Typically, letters of credit are opened for a period to cover production and shipping, and they are normally paid within seven working days of receipt of the goods. Cash payment or down-payments provide an extra measure of security for export sales.

B. Currency Controls and Domestic Credits

The establishment of a realistic Polish zloty exchange rate in 1989 and its internal convertibility as of January 1, 1990, represent steps towards external convertibility of the Polish currency. This facilitates trade. Companies operating in Poland have free access to foreign currency to finance imports, but they must exchange their hard currency export earnings which are held in zloty accounts. Though much opposed by business when introduced, this system has not proven problematic (owing to the stability of the zloty). There have been no failures within the banking system to provide hard currency on demand. Zloty profit repatriation by joint venture or wholly-owned foreign firms had been restricted, but a new foreign investment law allowing full repatriation of profits was passed in June 1991.

Companies may utilize zloty earnings to buy foreign currency to pay for imports. These measures have substantially opened the market for Western goods. Transfer of hard currency abroad, other than as payment for trade transactions, is strictly controlled. Domestic credit policy consistent with economic austerity is extremely tight. Interest rates (upwards of 5 percent per month) have remained prohibitively high during 1991.

The National Bank of Poland (NBP) acts as the central bank. There are 14 commercial banks empowered to conduct foreign exchange operations and to engage in the financing of foreign trade. The following five banks have been approved by the EximBank for bank guarantees, and currently engage in foreign trade financing. These banks maintain correspondent relationships with more than 2000 banks world-wide.

1. Bank Handlowy w Warszawie SA, headquartered in Warsaw, with branches in major Polish cities and in Western Europe;
2. Pekao SA, headquartered in Warsaw, with branches world-wide, including New York and Chicago;
3. The Export Development Bank, Bank Rozwoju Eksportu, Warsaw;
4. The State Savings Bank, PKO-BP, Warsaw;
5. The Bank for Food Economy, Bank Gospodarki Zywnosciowej, Warsaw.

The commercial banking network in Poland consists of 34 state-owned, cooperative and private banks. "The American Bank in Poland," the first foreign bank established in Poland since 1939, was licensed in 1990 and is a joint venture backed by a group of American corporations. Citibank and American Express have been recently licensed for operation and have full service offices in Warsaw.

There are currently 15 representative offices of Western commercial banks in Poland. Negotiations between several international banks and the Polish Ministry of Finance are ongoing and, though handicapped by problems emanating from the overall debt situation, the outlook for new developments in foreign banking is good.

Major Western European corporations tend to view recent developments in Eastern Europe with greater optimism. They have been more willing than U.S. corporations to overcome financing obstacles by extending direct supplier credits. In some cases availability of supplier credits may be a decisivee factor in the award of an import contract.

C. Sources of Financing

Sources of financing for projects and exports to Poland vary depending upon specific financial requirements and needs. The Export-Import Bank of the United States (EximBank) provides direct loans to Polish government or private sector buyers of U.S. products, loans to financial intermediaries, and guarantees to lenders to Polish buyers. Its affiliated agent, the Foreign Credit Insurance

Association (FCIA), assists U.S. exporters shipping on short and medium term credits by insuring nonpayment of Polish buyers for commercial and political reasons.

EximBank currently has confirmed the following Polish financial institutions as obligers or guarantors:

1) Bank Handlowy w. Warszawie S.A.,
2) Bank Polska Kasa Opieki S.A. (Pekao, S.A.),
3) Bank Rozwoju Eksportu S.A. (Export Development Bank),
4) Powszechna Kasa Oszczednosci-Bank Panstwowy (PKO-BP),
5) Bank Gospodarki Zywnosciowej (Bank for Food Economy).

There are no discretionary credit limits for short-term comprehensive FCIA insurance. Coverage is generally limited to irrevocable letters of credit issued by or guaranteed by the above mentioned banks. Other transactions are examined on a case-by-case basis. FCIA will also consider coverage of short-term comprehensive, short term political, and medium term insurance upon application for a Special Buyer Credit Limit.

The Overseas Private Investment Corporation (OPIC) is a self-sustaining U.S. government agency which provides investment information, financing, and political risk insurance for U.S. companies investing in developing countries including Poland. OPIC offers medium to long term financing in Poland through its direct loans and guarantees program. Direct loans are reserved for small U.S. companies involved in projects of $500,000 to $6 million. All-risk loan guarantees are issued to U.S. lending institutions. OPIC can insure U.S. investments against political violence, expropriation, and incovertibility of local currency. OPIC also has initiated a $200 million Central and East European Growth Fund (managed by Salomon Brothers) in which Poland participates. The Fund is designed to provide capital for new projects, expansion of existing also is developing an Environmental Fund of $100 million, to provide capital to U.S. companies involved in projects linked to economic development and the protection of the environment in Poland. For more information contact OPIC in Washington at (202) 457-7087.

The Polish-American Enterprise Fund was established by the U.S. Congress in 1989 to promote the development of the Polish private sector. The Enterprise Fund provides loans, grants, insurance, and guarantees to a variety of small to medium sized Polish businesses, agriculture concerns, and U.S.-Polish joint ventures. The fund was initially capitalized with $240 million from the U.S. Government. The Fund will consider investment in individual projects at levels up to 10 percent of its committed capital. It has made a total of $95 million in commitments in its first 13 months of activity. For more information contact 339-8330, or in Warsaw at 625-1921.

The World Bank is a source for long-term financing of projects in Poland. The World Bank traditionally finances infrastructural projects such as roads and power facilities. More recently, it has provided structural adjustment loans to help countries grow out of their financial problems through market-oriented policy and institutional reforms. The "International Bank for Reconstruction and Development" (IBRD) of the World Bank makes loans at market rates of interest to countries including Poland. The World Bank's "International Finance Corporation" provides loans and equity capital direct to private business. Established World Bank lines of credit in Poland include: $300 million for economic restructuring, $260 million for export-oriented projects, $250 million for gas exploration and production, $145 million for railway and road networks, $100 million for agro-industry, and $18 million for environment protection management. Lines under consideration include: $150 million for banking systems, $150 million for telecommunications modernization, and $100 million for the housing sector. Further information on World Bank programs can be obtained by calling the World Bank in Washington at (202) 477-1234, or in Warsaw at 635-0553.

The U.S. Trade Development Program (TDP) promotes U.S. exports through direst assistance to developing countries, including Poland. TDP accomplishes this by financing feasibility studies and related planning services and training programs. These projects must, as a rule, offer good export opportunities for U.S. exports of equipment and services. TDP has two avenues of feasibility study financing; one to the public sector, and the other to the private sector. The public sector "window" comprises more than 90% of TDP financing. It offers grants to U.S. firms in large scale public sector projects. The private sector "window" offers cost-sharing of feasibility studies with U.S. investors on a reimbursable basis. For further information contact the TDP in Washington at (703) 875-4818.

The U.S. Agency for International Development (AID) is now providing loans to the Polish government to allow Polish public and private sector importers to purchase U.S. products and services, including commodities and machinery. For information contact AID in Washington at (703) 875-4818, or at the American Embassy in Warsaw at 628-3041, extension 2292.

U.S. exporters should contact the U.S. Small Business Administration (SBA) for information on their Export Revolving Line of Credit, and other available assistance programs. To get the number of your local SBA small business development center, contact SBA in Washington at (202) 653-7561.

V. BARRIERS AND OBSTACLES TO TRADE AND INVESTMENT

There are problems in doing business in Poland today. International businesses currently face underdeveloped communications, banking, accounting, and distribution systems in Poland. Those companies establishing branch offices find offices and housing in short supply and very expensive. Communicating domestically and globally is a difficult and frustratingly slow procedure. International mail takes three, often four weeks. Domestic letters can take ten days between Polish cities. Telephones are in short supply, and domestic hookups for businesses sometimes take one to two years to arrange. Line connections are of poor quality and, though fax machines are available, fax transmissions suffer from the poor telephone connections. The recent introduction and expansion of modern U.S. and foreign telecommunication equipment and systems in Poland has helped to alleviate some of these problems.

Though the road and rail network in Poland is relatively extensive, and Poland's air and sea ports are structurally adequate for receiving and shipping cargo, all are in need of expansion and modernization to handle the continued growth of Poland's economy. Airport cargo modernization is underway, and modernization of the rail system is under examination.

Rural road travel is particularly dangerous at nigh. There are a lack of adequate (four-lane) highways between major cities which are capable of carrying the increased volume of trucks necessary to the growth of Poland's overall distribution system. Even in the most basic sense, getting the product to market is often a slow and arduous task by Western standards.

As mentioned earlier, the Polish zloty is not yet fully convertible. There is a relatively stable exchange rate internally, but the zloty is not expected to be fully convertible until 1996.

Western banking and accounting services are still in their infancy in Poland. Poland's banking system was recently decentralized, and interest rate policy was relaxed. Banks now set their own lending and deposit rates. In general, however, banking and accounting services are cumbersome, inefficient and a hindrance to business. Interest rates on loans to Polish businesses are very high. For all intensive purposes a checking system does not yet exist, and international wire transfers take time. The U.S. banks and accounting firms that currently operate in Warsaw have greatly helped to alleviate these problems, and Poland's own banking industry is slowly improving.

Finally, business in Poland suffers from a lack Western management training and experience. Former centrally controlled and managed enterprises are slow to adapt to modern business thinking and technique. Skilled labor is abundantly available, however; it is only a matter of time (and training) before Poland is technically capable of operating its market economy on par with the West.

A. Issues and Obstacles to Investment

The "Comprehensive Business and Economic Treaty," signed by President Bush and Poland's former Prime Minister Mazowiecki in March 1990, was ratified by Poland's parliament and signed by President Walesa in June 1991. Its purpose is to promote greater economic cooperation between the U.S. and Poland and foster long-term business and economic ties between the two nations. The treaty however does not come into force until its instruments of ratification have been exchanged.

New laws on privatization and foreign investment have been introduced in Poland. Amendments to the Law of Foreign Investments have reduced bureaucracy and government interference, modified tax incentives, and terminated restrictions on profit repatriation. Past needs for permission to establish operations, representative offices and joint ventures with Polish private partners have been reduced or eliminated. Where a permit is still necessary, the application process has been simplified.

Protection of intellectual property continues to be a main bilateral concern; the lack of it could be considered the overriding obstacle to increased foreign investment in Poland. The U.S. government continues to press the issue since adequate protection of intellectual property is called for under the Comprehensive Business and Economic Treaty. Poland's leaders seem to recognize the severe deficiency in intellectual property protection and its consequences. However, it has taken longer than expected to amen existing Polish laws to allow the application of international standards.

The U.S. government continues to encourage transparency in Poland's bid tendering procedure. Legislation governing the whole of this process is expected to be finalized in 1991, or early 1992. There is evidence that the Polish government is striving to tender bids fairly and transparently, but the lack of clear legislation and resulting delays continue to frustrate U.S. business. Allegations of unfair play occasionally surface.

In some cases, bureaucracy looms over Poland's state-industry privatization process. Projects and contracts have been reopened for examination and expanded bids. Inter-governmental committees have been set up to analyze and review major projects and consider bidders and proposals (usually adding an additional layer of consultants and investment bankers to the project). This often causes considerable delay in projects where the Polish company and U.S. company have already agreed to terms. The U.S. government continues to support U.S. companies in such instances and continues to push for U.S. participation in Polish major projects.

B. Issues and Obstacles to Trade

The previous 3-year exemption from customs duty on imports necessary to conduct business activity has been dissolved under the new Law on Foreign Investments. Imported fixed assets (in-kind capital) are exempted from customs duty, provided they are not disposed of for three years. In January 1990, a new Customs Law came into force in Poland. This law defined and implemented the Harmonized Tariff System which reclassified products and ultimately formed the basis for new duty rates on imports. The average import tariff in Poland is now about 12 percent, with duty rates ranging from 0 to 45 percent and those on raw materials and processed products in the range of 0 to 10 percent. Some consumer goods now have higher duties. U.S. exporters accordingly have shown frustration with these rates, exporters accordingly have shown frustration with these rates, and with what seems to be a lack of communication of Polish Government tariff policy to customs officials at the borders.

Taxes can add up when doing business in Poland. Imports into Poland are subject to a turnover tax that averages 20 percent of the cost of the goods including their duty. This tax is higher on certain consumer products like cars, alcohol and tobacco. Poland is expected to replace the turnover tax with a Value Added Tax (VAT) in 1992.

All companies operating in Poland are subject to Polish law. Currently, a company must pay

20% tax on wages paid to Polish employees. All wages of U.S. employees must be paid in zloty, from which the company must also withhold 20 percent. (A personal income tax law, to be introduced in January 1992, will change the wage taxes.) In addition to the tax on wages, a company must also pay 45% of the gross wage of both Polish and U.S. employees to unemployment insurance and the Polish Social Security Fund. The U.S. government, at the request of the American Chamber of Commerce in Poland, is currently examining relief options for U.S. expatriot employees in Poland on limited assignments.

The corporate income tax in Poland is 40% of taxable profits. The United States has signed a double taxation treaty with Poland, and it enjoys a reduced rate of withholding tax on dividends. In corporations where a U.S. firm holds a minimum of 10% of the equity, there is a 5% tax on dividends. In all other cased it is 15 percent.

With few exception, import licenses are no longer required in Poland. However, certain high-technology imports into Poland may still require a U.S. export license. A Polish import certificate is necessary to support the U.S. license.

VI. MARKET ANALYSIS PLAN

Following are topical titles of marketing information reports, and their scheduled dates of completion. Where appropriate, additional detail is included (e.g. names of contacts for reports, descriptions).

A. Industry Sector Analyses

1. Air Pollution Measuring Equipment (POL)
Country: Poland
Expected Completion Date: November 1991
Key Researcher: Subcontractor
Estimated Time of Research (Days): 90
Justification: High Priority Sector
Costs Incurred: $3500

2. Personal Computer Software and Services (CSF)
Country: Poland
Expected Completion Date: February 1992
Key Researcher: Maciej Kaseja, FCS Warsaw
Estimated Time of Research (Days): 90
Justification: High Potential for U.S. Exports
Approximate Costs: $1000

3. Consumer Health Care Products (HCS)
Country: Poland
Expected Completion Date: March 1992
Key Researcher: Marysia Kowalska, FCS Warsaw
Estimated Time of Research (Days): 90
Justification: High Potential of U.S. Exports
Approximate Costs: $1000

4. Hotel and Restaurant Equipment (HTL)
Country: Poland
Expected Completion Date: March 1992
Key Researcher: Ewa Kostecka, FCS Warsaw

Estimated Time of Research (Days): 90
Justification: High Potential for U.S. Exports
Approximate Costs: $1000

5. Housing Construction Equipment (CON, BLD)
Country: Poland
Expected Completion Date: March 1992
Key Researcher: Basia Grabowska, FCS Warsaw
Estimated Time of Research (Days): 90
Justification: High Potential for U.S. Exports
Approximate Costs: $1000

6. Sporting Goods and Equipment (SPT)
Country: Poland
Expected Completion Date: March 1992
Key Researcher: Aleksandra Prus, FCS Warsaw
Estimated Time of Research (Days): 90
Justification: High Potential for U.S. Exports
Approximate Costs: $1000

7. New and Used Passenger Cars (AUT)
Country: Poland
Expected Completion Date: May 1992
Key Researcher: Maciej Kaseja, FCS Warsaw
Estimated Time of Research (Days): 90
Justification: High Potential for U.S. Exports
Approximate Costs: $1000

8. Do-It-Yourself Home Tools and Hardware (BLD)
Country: Poland
Expected Completion Date: July 1992
Key Researcher: Aleksandra Prus, FCS Warsaw
Estimated Time of Research (Days): 90
Justification: High Potential for U.S. Exports
Approximate Costs: $1000

9. Office Furniture, Equipment, and Supplies (FUR)
Country: Poland
Expected Completion Date: July 1992
Key Researcher: Ewa Kostecka, FCS Warsaw
Estimated Time of Research (Days): 90
Justification: High Potential for U.S. Exports
Approximate Costs: $1000

10. Architectural and Engineering Services (ACE)
Country: Poland
Expected Completion Date: September 1992
Key Researcher: Ewa Kostecka, FCS Warsaw
Estimated Time of Research (Days): 90
Justification: High Potential for U.S. Exports
Approximate Costs: $1000

11. Industrial Pollution Control Equipment (POL)
Country: Poland
Expected Completion Date: September 1992
Key Researcher: Maciej Kaseja, FCS Warsaw
Estimated Time of Research (Days): 90
Justification: High Potential for U.S. Exports
Approximate Costs: $1500

B. Supported Trade Event Market Packages

USFCS Warsaw will support marketing packages for existing trade promotion events and upcoming trade missions. For additional information on trade events in Poland, contact the East European Business Information Center (EEBIC) at (202) 377-2645.

Scheduled Event Market Packages

1. Subject: Industrial and Consumer Goods Market
   Event: Poznan International Fair, June 1992
   Completion: December 1991
2. Subject: Food Processing/Agro-Industrial Market
   Event: Polagra Trade Fair, October 1992
   Completion: February 1992

C. Scheduled Periodic Reporting

Title of Report	Date Available
Annual Listing of Holidays	November 1991
Annual Listing of Trade Fairs	November 1991
Foreign Economic Trends Report	June 1992
Market Share Reports	June 1992

D. Other Post-Initiated Market Insight Reports

USFCS Warsaw expects to generate a series a topical reports in the coming fiscal year, which will correspond to evolving developments in the business community in Poland. Market insight reports will be brief and submitted on a timely and relevant basis. They will be a significant part of our newly revised "Business Information Kit" which will be made available to U.S. exporters though the Eastern European Business Information Center (EEBIC), or to those who visit our office in Warsaw.

VII. TRADE EVENT PLAN

A.	U.S. Exhibition at the Polagra Trade Fair
Date:	Oct. 3-8, 1991
Theme:	Food Processing/Agro-Industry
Type of Event:	Trade Fair
Location:	Poznan, Poland
Washington Contact:	Alain de Sarran (202) 377-1599
Participants:	Food processing/agricultural firms experienced in other export markets, but new to Poland.

B. Mining Equipment Trade Mission

Date: Oct. 20-24, 1991
Theme: Mining Industry
Type of Event: IOGA Trade Mission
Location: Warsaw, Katowice, Poland
Washington Contact: Audrey Zuck (202) 377-4915
Participants: Exploration, mine development, operation, handling, preparations and processing.

C. National Machine Tool Builders Association Trade Mission

Date: March 1-3, 1991
Theme: Machine Tool Industry Association
Type of Event: IOGA Trade Mission
Location: Warsaw, Poland
Washington Contact: Richard Kuba (703) 893-1151
Participants: Key officers of the association

D. Environmental Systems Trade and Investment Mission

Date: March 11-14, 1991
Theme: Environmental Control Technology
Type of Event: IOGA Trade Mission
Location: Warsaw, Katowice, Poland
Washington Contact: Damon Greer (202) 377-0564
Participants: Systems manufacturers, environmental engineers, designers and consulting firms.

E. U.S. Exhibition at the Poznan International Fair

Date: June 7-14, 1992
Theme: All Industry
Type of Event: Trade Fair
Location: Poznan, Poland
Washington Contact: Alain de Sarran (202) 377-1599
Participants: Every sector, old and new to market. Poland's largest trade show.

F. Scheduled Catalog Shows

Topic of Show	Season
High Tech USA	Spring 1992
Food Production	Fall 1992
Pollution Control	Fall 1992

G. Foreign Buyer Program Events

USFCS Warsaw will support all Foreign Buyer Program events that relate to Best Prospects for U.S. exports to Poland.

H. 1992 Polish Holidays

Day, Date	Holiday
Wednesday, January 1	New Year's Day
Monday, April 20	Easter Monday
Friday, May 1	Labor Day
Sunday, May 3	Constitution Day
Thursday, June 18	Corpus Christi Day
Saturday, August 15	Assumption of the Virgin Mary
Sunday, November 1	All Saints Day
Wednesday, November 11	All Saints Day
Friday, December 25	Christmas Day
Saturday, December 26	Boxing Day

APPENDIX K - MARKETING PLAN: HUNGARY

This appendix was made available by the U.S. and Foreign Commercial Service, Budapest, Hungary. It is dated November 1991.

I. COUNTRY DATA

A. Profile

Population:	10.35 Million
Religion:	Catholic-69%; Calvanist-20%; Lutheran-5%; Other-8%.
Government:	Multi-party parliamentary democracy. Coalition government formed in May 1990. Prime Minister: Jozsef Antall (Hungarian Democratic Forum). President: Arpad Goncz (Alliance of Free Democrats).
Language:	99% Hungarian; German and English typically used in business discussions.
Work Week:	40 hours, Monday-Friday

Contacts

Name	Phone	Fax	Location
Gary L. Gallagher Commercial Attache	(36)(1) 122-8600	(36)(1) 142-2529	American Embassy-USFCS Bajza utca 31 1062 Budapest, Hungary
(mailing address):			c/o USFCS AmEmbassy (Budapest) APO NY, 09213-5270
Russell Johnson Desk Officer for Hungary	(202) 377-2645	(202) 377-2155	IEP/OEESA/EED-Rm 3414 U.S. Dept. of Commerce Washington, DC 20230
George Knowles Regional Manager	(202) 377-1599	(202) 377-3159	US&FCS/OIO-Room 3122 U.S. Dept. of Commerce Washington, DC 20230
Hans Amrhein Regional Development Officer	(202) 377-1599	(202) 377-3159	US&FCS/OIO-Room 3122 U.S. Dept. of Commerce Washington, DC 20230
Susan Blackman East European Business Info. Center	(202) 377-2645	(202) 377-4473	EEBIC-Room 6043 U.S. Dept. of Commerce Washington, DC 20230

B. Domestic Economy

(Current billion US$)	1989	1990	1991(E)
GDP*	28.87	27.72	26.33
Real GDP Growth Rate (%)	-0.20	-4.30	-5.00
GDP Per Capita (US$)	2,645	2,364	2,328
Government Spending (% GDP)	65.20	60.70	56.00
Inflation (%)	17.00	29.00	31.00
Unemployment (%)	0.60	1.60	3.00
Foreign Exchange Reserves (excl gold)	1.25	1.07	1.80
Average Exchange Rate ($1=HUF)	62.54	62.99	75.85
Foreign Debt/Gross (Mil US$)	20.97	21.27	21.70
Debt Service Ratio	48.40	45.50	40.00
U.S. Economic Assistance	0.00	32.50*	NA
U.S. Military Assistance	0.00	0.00	0.00

C. Trade	1989	1990	1991(E)
Total Exports (Bil US$)	10.50	9.05	14.03
Convertible Currency	6.45	6.35	14.03
Non-Convert Currency	4.05	2.70	—**
Total Imports (Bil US$)	9.45	8.50	11.90
Convertible Currency	5.91	6.00	11.90
Non-Convert Currency	3.54	2.50	—**
U.S. Exports (Mil US$)			
Hungarian Statistics	92.70	338.80	NA
U.S. Statistics	122.00	156.40	221.30
U.S. Imports (Mil US$)			
Hungarian Statistics	305.00	228.60	NA
U.S. Statistics	328.00	347.60	374.30
U.S. Share Host-Country Imports (%)	1.60	2.87	3.10
Total Manufactured Imports (Bil US$)	4.74	5.90	8.26
U.S. Share of Mfr Imports (%)	1.30	1.54	1.80

1990 Trade Balance with 3 Leading (Convertible Currency) Partners:	Exports	(Million US$) Imports	Balance
1. Germany	1916.90	2017.10	-100.20
2. Austria	719.10	860.60	-141.50
3. Italy	562.80	350.40	121.40

---

* Hungary received approximately $87.5 million from SEED-1 —$60 million is for the Hungarian-American Enterprise Fund which is being distributed over a 3-year period: $5m in 1990; $25m in 1991; and, $30m in 1992. SEED-2 is likely result in additional assistance.

** As of January 1, 1991, all trade is conducted on a dollar-denominated basis.

Principal U.S. Exports: (FAS, Million US$)		**1990**
HS 100510	Corn (maize) sd, certif., excl. sweet corn	26.30
HS 381600	Refractory cmnts, mortars, concretes, nesoi	6.90
HS 870190	Tractors, nesoi	4.40
HS 870860	Road wheels & parts for motor vehicles	3.40
HS 700239	Tubes of glass, nesoi, unworked	2.80
HS 240399	Mfr tobacco & substitutes nesoi	2.60
HS 880211	Helicopters of an unladen weight < 2000 kg	2.60
HS 852490	Recorded media for sound incl mast1er prod	2.50

Principal U.S. General Imports: (Customs Value, Million US$)		**1990**
HS 870850	Drive axles with differential for motor vehicles	28.10
HS 200970	Apple juice, unfermented, sweetened or not	23.70
HS 160241	Prepared or preserved swine nesoi, hams,	22.20
HS 640399	Ftwr, outer sole rubber, & leather upper nesoi	20.10
HS 870860	Non-driving axles & parts for motor vehicles	15.00
HS 401120	New pneumatic tires, rubber, for bus/truck	13.20
HS 160242	Prep or preserved swine, nesoi, shoulders	8.00
HS 294190	Antibiotics, nesoi	5.70

D. Investment (Million US$)	**1989**	**1990**	**1991(E)**
Total Direct Foreign Investment	400	1,200	3,000
U.S. Investment	N/A	750	1,500

Principle Foreign Investors: United States, Germany, Austria. (Note: Germany, Austria and Switzerland participate in more joint ventures than the United States. However, the United States is the principle supplier of capital).

Sources:
Annual Report, National Bank of Hungary, 1990;
International Financial Statistics, IMF, April 1991;
Market Letter, National Bank of Hungary, 3/1991;
Official Statistics of the U.S. Department of Commerce;
Quarterly Review, National Bank of Hungary, 2/1991;
Statistical Yearbook of External Trade-1989, Hungarian Statistical Office;
Statistical Yearbook-1989, Hungarian Statistical Office;

II. BEST PROSPECTS

[Note: Statistical data in Hungary is difficult to find. Estimates presented below were often developed based on USFCS/Budapest's market knowledge, discussions with industry and government bureaus, and information obtained from printed articles and journals. Estimates should not be considered "hard data." In many instances, the dearth of statistical information obviated the development of a credible estimate.

Emphasis should be placed on the narrative descriptions. In general, current market size and U.S. sales data belie the potential of the Hungarian market to U.S. businesses. The Best Prospects List underscores areas where market demand is growing.]

A. Rank of Sector: 1
B. Name of Sector: Food Processing and Packaging Equipment
C. Code: FPP
D. Est. 1991 Market Size (USDOL Millions): $100
E. Est. Average Annual Growth Rate of Market '91-'93: 3-4%
F. Est. 1991 Total Imports (USDOL Millions): N/A
G. Est. Average Annual Growth Rate of Total Imports: N/A
H. Est. 1991 Imports from the U.S. (USDOL Millions): $1-2
I. Est. Average Annual Growth Rate of Imports from the U.S.: 2%
J. Score of Country Receptivity to U.S. Products (5=Highest, 0=Lowest) 4
K. Subsectors: Processing equipment for: Meat/Poultry; Fruit/Vegetable Canning; Dairy; Fats/oils. Packaging equipment.
L. Description: Agricultural and processed foods generate significant hard currency export earnings for Hungary. Government intention to increase exports by 20 percent, with emphasis on processed foods (unduly high share of agricultural products exported with little processing or as unprocessed raw material). Productivity in processed food sector is low. Installed equipment is over-aged and outdated (some modernization in recent years in meat/poultry segment only). Packaging is a weal point for Hungarian food processors. Likely that demand for packaging equipment will increase faster than for processing equipment. Hungarian companies are looking to form joint ventures with Western equipment makers. Amended Investment Act maintains the 5-year tax holiday for foreign investment in this sector.

A. Rank of Sector: 2
B. Name of Sector: Telecommunications Equipment
C. Code: TEL
D. Est. 1991 Market Size (USDOL Millions): N/A See Description
E. Est. Average Annual Growth Rate of Market '91-'93: N/A
F. Est. 1991 Total Imports (USDOL Millions): N/A
G. Est. Average Annual Growth Rate of Total Imports: N/A
H. Est. 1991 Imports from the U.S. (USDOL Millions): N/A
I. Est. Average Annual Growth Rate of Imports from the U.S.: N/A
J. Score of Country Receptivity to U.S. Products (5=Highest, 0=Lowest) 5
K. Subsectors: Telephone systems; switching apparatus; telephones (portable, cellular, etc.); fax machines; answering machines; parts, etc.
L. Description: Hungary has one of Europe's least developed telecommunications systems. Government priority to upgrade system—modern telecommunications infrastructure viewed as vital in order to compete internationally and to attract Western investment capital in other sectors. Currently, only 39 percent of Budapest households have a telephone: outside Budapest, this figure drops to 10 percent. The waiting list for phones stands at 600,000 (waiting time=13 years). As of 1988, Hungary had an installed base of 1,048,924 main telephone lines. The telex network had 12,614 subscribers. World Bank has provided financing for modernization of telecom system. In 1987, the World Bank granted $70 million loan ($45 million has been dispursed so far). In October 1990, World Bank approved a $150 million loan agreement to finance the Second Telecommunications Development Program which will be completed by

1995. In November 1989, Hungarian Cellular Communications (US West and Hungarian Telecommunications Co. (HTC)) won tender to install 450 MHz cellular system. Two additional tenders for a 900 MHz GSM system are currently being disputed in the Hungarian courts. Ericsson and Siemens won a digital switching tender in December 1990. The Government's ten-year development program for public switching network will cost $6.3 billion. Upgrade will include: 3 million additional phone lines; data transmission capability; digital transmission and switching apparatus; optical fiber transmission; integrated services digital network (ISDN).

A. Rank of Sector: 3
B. Name of Sector: Tourism
C. Code: TRM
D. Est. 1991 Market Size (USDOL Billions): $2.6
E. Est. Average Annual Growth Rate of Market '91-'93: 11%
F. Est. 1991 Total Sales of Foreign-Owned Firms in 1991: N/A
G. Est. Average Annual Growth Rate of Foreign Firms Sales: N/A
H. Est. Total Sales of U.S. Firms 1990: N/A
I. Est. Average Annual Growth Rate of U.S. Sales: N/A
J. Score of Country Receptivity to U.S. Products (5=Highest, 0=Lowest) 5
K. Subsectors: Hotel/resort development; travel services; retail; amusement/entertainment.
L. Description: Outside of the Budapest and Lake Balaton tourist areas, many regions remain undeveloped for tourism. Hungary hosted record amounts of visitors in 1990—37.6 million people, an increase of 51 percent over 1989. Of this total, 17.4 million came from Western countries. The Government is projecting that visitors will increase to 43 million in 1994. Most visitors to Hungary were from Romania and Yugoslavia—the European and American markets are yet untapped. For Europeans, Hungary can be particularly enticing. Hungary has been a favorite, inexpensive get-away know to Germans and Austrians for some time. Tapping the mainstream European/American tourist markets means bringing in tourists with, on average, higher disposable (and discretionary) incomes that demand higher quality accommodations, food, etc. Of the total foreign direct investment in Hungary, approximately 10 percent of this total is related to tourism. Opportunities exist in hotel development particularly—Hungary is experiencing a shortage of hotel rooms. Five-year tax holidays are provided for foreign ventures in hotel construction.

A. Rank of Sector: 4
B. Name of Sector: Computers and Peripherals
C. Code: CPT
D. Est. 1991 Market Size (USDOL Millions): N/A
E. Est. Average Annual Growth Rate of Market '91-'93: 3%
F. Est. 1991 Total Imports (USDOL Millions): N/A
G. Est. Average Annual Growth Rate of Total Imports: N/A
H. Est. 1991 Imports from the U.S. (USDOL Millions): N/A
I. Est. Average Annual Growth Rate of Imports from the U.S.: N/A
J. Score of Country Receptivity to U.S. Products (0=Lowest; 5=Highest) 5
K. Subsectors: PCs, mainframes, components, printers, monitors
L. Description: There is a substantial demand in Hungary to upgrade the information technology.

Computers are viewed positively in augmenting productivity and increasing the competitiveness of existing and new enterprises. Hungary has had little access in the past to Western high tech products. Until recently, export control regulations were stringent with respect to East bloc countries. Production of computers started 5 years ago—50% of the market supplied by imports. Three principle manufactures in Hungary: Vidoton, Controll and Muszertechnika. Most computers produced in Hungary are assembled from imported components from the Far East. PCs mainly compatible with IBM PC/AT and PC/XT. Hungary can obtain Intel 80286, 80386 and 80486 microprocessors. Installed PC base is approximately 250,000 (97% IBM compatible). Installed mainframe base is 600 (50% from West—IBM, Siemens, Bull-Honeywell, ICC, CDC). Installed minicomputer base is approximately 500. Total value of installed base is $1 billion. While the Hungarian market is not saturated it is maturing rapidly. All major American computer manufacturers are active in Hungary (IBM, Compaq, DEC, Hewlett-Packard) besides in addition to the major European producers.

A. Rank of Sector: 5
B. Name of Sector: Construction Equipment
C. Code: CON
D. Est. 1991 Market Size (USDOL Millions): N/A
E. Est. Average Annual Growth Rate of Market '91-'93: N/A
F. Est. 1991 Total Imports (USDOL Millions): $35
G. Est. Average Annual Growth Rate of Total Imports: 7%
H. Est. 1991 Imports from the U.S. (USDOL Millions): $0.5
I. Est. Average Annual Growth Rate of Imports from the U.S.: 1%
J. Score of Country Receptivity to U.S. Products
(5=Highest, 0=Lowest) 5
K. Subsectors: Earth moving machinery, cranes, dump trucks, etc.
L. Description: The construction industry will play a major role literally, in re-building the economy. The Government has plans to modernize and expand the transport systems. Authorities have been consulting with Western countries on augmenting housing and office space which is in short supply. U.S. construction equipment firms, unlike their West European competitors, are internationally competitive. With trade in capital goods switching from CMEA to the West, U.S. producers will benefit. Some U.S. construction service firms are already present in the Hungarian market.

A. Rank of Sector: 6
B. Name of Sector: Medical Equipment
C. Code: MED
D. Est. 1991 Market Size (USDOL Millions): $50
E. Est. Average Annual Growth Rate of Market '91-'93: 0%
F. Est. 1991 Total Imports (USDOL Millions): $25
G. Est. Average Annual Growth Rate of Total Imports: -2%
H. Est. 1991 Imports from the U.S. (USDOL Millions): $3-4
I. Est. Average Annual Growth Rate of Imports from the U.S.: 2%
J. Score of Country Receptivity to U.S. Products
(5=Highest, 0=Lowest) 4
K. Subsectors: Electro-medical equipment—x-ray based equipment, CAT scans, etc.
L. Description: Healthcare is controlled by the Government. As yet, there are no private hospitals although legislation is expected. Equipment in the healthcare system is dated. However, because of domestic budgetary concerns, the Government is unlikely to divert the necessary funds to

upgrade the system. In fact, the market is likely to constrict during 1990-91. As the Government reforms management of the system, opportunities will be created for American businesses. particularly with U.S. firms advanced technology. Currently, the electro-medical equipment market is approximately $50 million. Major foreign suppliers are Germany and Austria.

A. Rank of Sector: 7
B. Name of Sector: Pollution Control Equipment
C. Code: POL
D. Est. 1991 Market Size (USDOL Millions): N/A
E. Est. Average Annual Growth Rate of Market '91-'93: N/A
F. Est. 1991 Total Imports (USDOL Millions): N/A
G. Est. Average Annual Growth Rate of Total Imports: N/A
H. Est. 1991 Imports from the U.S. (USDOL Millions): N/A
I. Est. Average Annual Growth Rate of Imports from the U.S.: N/A
J. Score of Country Receptivity to U.S. Products (5=Highest, 0=Lowest) 4
K. Subsectors: air/water pollution control equipment; pollution clean-up equipment.
L. Description: The former regime's commitment to pursue industrialization at all costs has resulted in an enormous environmental bill. Hungary's environmental problem is not as severe as its neighbors. Local and municipal governments have made pollution control a priority. For example, in Budapest, the air pollution control could adversely affect the tourism industry. Tenders gradually are being issued. However, opportunities may still be further in the future. While the environment is a priority, so too is development. Pollution regulation could conflict with this latter objective.

A. Rank of Sector: 8
B. Name of Sector: Household Consumer Goods
C. Code: HCG
D. Est. 1991 Market Size (USDOL Billions): $10
E. Est. Average Annual Growth Rate of Market '91-'93: 0.3%
F. Est. 1991 Total Imports (USDOL Billions): $2
G. Est. Average Annual Growth Rate of Total Imports: 4-7%
H. Est. 1991 Imports from the U.S. (USDOL Millions): $120
I. Est. Average Annual Growth Rate of Imports from the U.S.: 1.9%
J. Score of Country Receptivity to U.S. Products (5=Highest, 0=Lowest) 4
K. Subsectors: paper, furniture, etc.
L. Description: Hungary has a larger consumer goods industry than most East European countries. For example, Herend china is exported and glassware is internationally price competitive. However, other consumer goods segments do not achieve the quality of Western-manufactured products. There is a pent-up demand for products previously in short supply: consumer and household appliances. A number of U.S. consumer products companies are investing in Hungary: Samsonite, Levi-Strauss, Schwinn, etc. Hungarian companies are actively seeking joint venture partners. The Government encourages foreign investment in this sector by offering generous tax incentives. U.S. consumer goods export opportunities are likely to lag improvements in the domestic economy. Currently, living standards are low—monthly salaries are approximately $200. Consequently, many families do not have large discretionary incomes to afford costly Western luxury items. The Government estimates that domestic production by

"light industry" fell 10 percent in real terms. Clothing (which accounts for approximately 60 percent), among other traditional consumer goods, is included in Hungarian light industry statistics. In 1990, clothing production decreased while furniture and paper output expanded. Exports of light goods dropped to Eastern Europe, but grew 18 percent to Western countries. Last year also saw the establishment of 300 new companies in this industry segment.

III. COMMERCIAL ENVIRONMENT

A. Political Factors

Hungary was a communist country for over forty years following WWII. In 1956, a populist revolt was suppressed by Soviet military intervention. The Government was purged and Janos Kadar, a hardliner, assumed the Party leadership for the next 32 years.

In 1988, a new government was ushered in, composed of more liberal communists, which undertook a number of reforms. In the Fall of 1989, the government opened the Austrian border allowing a flood of East German refugees to cross into the West. This action had a doplar-like effect across the Eastern Bloc, precipitating the implosion of the various communist governments. With the world's attention focused on events such as the destruction of the Berlin Wall and the Romanian street fighting, Hungary quietly underwent its own political transformation (revolution).

There are six major political parties in Hungary: the Hungarian Democratic Forum (MDF), the Alliance of Free Democrats (SZDSZ), the Young Democrats (FIDESZ), the Independent Smallholders (FKgP), the Christian Democrats (KDNP) and the Hungarian Socialist Party (MSZP). The free elections of April 8, 1991, gave 58 percent of the votes to the MDF. In May 1990, Joszef Antall of the MDF formed a coalition government with the SZDSZ and the KDNP which together account for approximately 60 percent of the 384 Parliamentary seats.

B. Economic Factors

Prelude to Reform: Hungary's economic development has been uneven. With Budapest as the eastern capital of the Austro-Hungarian empire, Hungary flourished. After WWI, the Treaty of Trianon redistributed formerly Hungarian-controlled territories. On the losing side of WWII, Hungary lost additional territories and suffered occupation by the Soviet army. Under communist rule after 1946, Soviet-style central planning proved counterproductive, plunging much of the nation into sub-standard living conditions. Mounting dissatisfaction led to the 1956 populist revolt which was quelled by Soviet intervention. Twelve years later, with no tangible improvements for Hungarian citizens, the Prague Spring threatened to have spillover effects. Responding to the worsening situation in order to eclipse and to head-off a potential crisis, the Communist authorities introduced limited market reforms (this experiment became known as "goulash communism")

In 1968, the Government of Hungary implemented the New Economic Mechanism (NEM) in order to mitigate mounting dissatisfaction. The NEM marked an official move away from strict central economic planning to limited reliance on market mechanisms. Yet, despite this revision in official thinking, Hungary's economy continued to be dominated by the State sector. According to the World Bank, Hungary experienced during the 1970/1980s chronic excess demand, stagnating output, low productivity and increased external indebtedness. Average annual GDP growth was only 1.5 percent while the national budget deficit doubled and Hungary acquired the largest per capita foreign debt in all of Eastern Europe. In 1987, the government implemented a three-year recovery program to contain convertible currency debt growth and to accelerate structural change. Simultaneously, the Government sought to introduce bankruptcy legislation, to decentralize the banking system and to reform the tax system with the creation of a personal tax and a VAT.

Economic Reform: In 1988, Janos Kadar, leader of the communist government since 1956, was replaced by Karoly Grosz as Party Chief and under him as Prime Minister was Miklos Nemeth. Under Nemeth's direction, newly-appointed officials pressed ahead with the enactment of reforms—austerity measures were implemented to reduce the country's trade deficit and hard currency debt, while the government took steps to encourage free enterprise. The freely elected, noncommunist coalition government (headed by Jozsef Antall) which was formed in May 1990, plans to intensify and accelerate the bureaucratic and structural changes initiated by Grosz.

Although the government continues to control credit and regulate foreign trade to some extent, price and wage controls have been significantly relaxed. Managers of state enterprises, which still account for 90 percent of national income, have essential control over developing their business strategies and making investment, production and pricing decisions. Competitive wholesale pricing has been introduced in most sectors and government subsidies are being eliminated gradually to increase competitiveness and to promote efficient resource allocation.

In September 1990, the government announced the "Economic Program for National Renewal," a three-year program which continues to emphasize privatization, import liberalization and deregulation, increased foreign investment, a curb on inflation, progress towards achieving currency convertibility, public finance reform, and welfare reform.

INVESTMENT AND PRIVATIZATION:

Legal Framework

The Government views foreign direct investment as playing a key role in economic development, modernization and improving competitiveness. Parliament has adopted enacted legislation to facilitate foreign investment. Three laws are particularly noteworthy:

> Act VI of 1988 on Economic Associations ("the Companies Act") defines and regulates the legal forms of business entities (excepted are cooperatives, sole traders and existing state businesses). Most foreign investment in Hungary takes the form of a limited company—"Korlatolt felelossegu tarsasag" or Kft—so as to limit liability and maintain the right to trade for profit. The Kft is similar to the German GmbH.
>
> Act XXIV of 1988 on Investments of Foreigners in Hungary ("the Investment Act") sets forth the rules of establishment and operations of joint ventures with foreign participation. A notable feature of the Act is that joint ventures enjoy national treatment. The Investment Act expands upon the Companies Act and tax laws favorable to foreigners, and provides additional incentives for foreigners. The Investment Act became effective January 1, 1989 and was amended by Parliament in January 1991. The amended version maintains the generous tax benefits for foreign ventures.
>
> Act XII of 1989 on Transformation of Economic Organizations and Economic Associations ("the Transformation Act") supplements the Companies Acts by providing the legal framework for state enterprises to be transformed into independent entities or new companies. The Transformation Act allows for foreign participation. The Act became effective,on July 1, 1989.

These laws have achieved their intended purpose—to encourage and to facilitate privatization and foreign investment. In 1990, 15,000 private enterprises emerged. Between January-June 1991, an additional 7,00 private companies were formed. In 1988, there were 250 joint ventures in Hungary. In June 1991, the government estimated there were 8,000 joint ventures operating, approximately 2,000 formed during the first two quarters of 1991. The government calculates that Hungary accounts for 50 percent of the capital invested in Eastern Europe.

Transformation of State Assets

The government plans to privatize 60 percent of all state-owned enterprises, which comprise 90 percent of all Hungarian enterprises. within 5 years. Well over 60 percent of all Hungarian businesses are not operating profitably. The government hopes to transfer managerial and technical skill to such enterprises through private transaction which may take the form of majority foreign-owned joint ventures, lease transactions or management contracts.

The 1990 law on the Protection of State Property and the creation of the state-supervised State Property Agency (SPA) provide the regulatory foundation for Hungary's privatization efforts. Since March 1990, the SPA has involved itself in about 200 privatization transactions. In September 1991, the SPA announced a list of 20 large state-owned enterprises to be privatized in the First Privatization Program. This list included: Richter Gedeon PT (pharmaceuticals), IBUSZ Travel, HungExpo and HungarHotels. At the end of 1990, SPA issued a preliminary list of enterprises to participate in the second round. A third privatization program is expected to be announced towards the end of 1991. This program will focus exclusively on the construction industry.

The ambivalent approach of the Government to the pace of privatization could threaten to dampen temporarily investor enthusiasm. For example, the sale of IBUSZ shares on the Budapest and Vienna Exchanges precipitated sharp criticism of the SPA. Earlier, the half-equity sale of HungarHotels to a Swedish concern was reversed retroactively because the price was considered too low. Underlying the Government's ambivalence is the distrust that state-owned assets will be acquired at "bargain-basement" prices by foreign investors. Yet, there is official recognition that foreign investment capital is essential to achieve restructuring objectives

Property Ownership

The Government continues to grapple with the issue of nationalized properties and whether to return the lands to former owners. In June 1991, the Parliament passed the Compensation Act which addresses how to compensate former owners of nationalized properties. This legislation is the first step in resolving the ownership issue. Parliament will now focus its attention on the land laws.

Companies with foreign ownership may buy and own property (including buildings, warehouses, factories, retail outlets, capital equipment and land) for conducting business. No property acquisition is permitted for speculative purposes, although this may change if certain proposed amendments to the land laws are adopted. State-owned property must be acquired with the approval of Executive Committees, and city and county councils. Securing ownership is not an automatic process. If actual ownership is denied, long-term property usage rights may nevertheless be obtained.

The Economic Outlook: Economic activity in 1990 was slower than expected due to a decrease in ruble exports and a weakening of domestic demand. Real GDP fell 4 percent in 1990 and declined throughout the first six months of 1991. A 5 percent drop in GDP is expected for all of 1991. Inflation weakened the economy in 1990, hitting nearly 30 percent. While inflation was 35 percent at mid-year, by the end of 1991 it is expected to be around 30 percent. This figure would represent an enormous achievement for the Government in preventing inflation from spiraling upwards. The structural realignment of industries has introduced the Western phenomenon of unemployment. In 1990, unemployment reached 1.6 percent. In 1991, unemployment is expected to break the 3 percent mark. Unemployment is an issue that is of will become of increasing concern to the Government.

Hungary's immediate outlook, thus far, has not been substantially affected by the effects of its modified trade relations with the CMEA countries. While these markets have "bottomed-out"—for example, exports to the USSR accounted for only 5 percent of total exports in the first quarter of 1991—Hungary's exports to the West continue to climb dramatically (up 30 percent at the close of Q1 1991). However, the combination of the drop in exports to the ruble zone and the sluggish domestic demand will continue to dampen industrial production. The implementation of the stabilization program and associated reforms is critical to Hungary's economic performance. Assuming the continuation of Hungary's strong export performance and tight domestic fiscal policies, Hungary could continue to register a positive current account, thereby reviving domestic industrial production.

Foreign Trade—Regulations: The decentralization of foreign trade activities is an important feature of the government's overall program. Hungary has achieved substantial progress in a short time. Foreign trade organizations, known as IMPEXs, no longer maintain exclusive rights to import and export products. Foreign trade may be undertaken by any economic organization or individual registered with the government. Three years ago, the government implemented a program to liberalize the import licensing regime. As of January 1991, 93 percent of imported products no longer require prior approval. Those that still require import licenses are primarily consumer goods which are subject to quotas. However, these quotas have not altered the trade flow, patterns—quotas have remained unfilled and the Ministry of International Economic Relations (MIER) has issued licenses in response to requests.

Foreign Trade—Performance: In 1990, Hungary recorded a record convertible trade surplus of $945 million, following a surplus of $540 million in 1989. The 1990 surplus resulted from dynamic export growth accompanied by only a modest increase in imports. In 1990, Hungary's hard currency exports grew 18 percent, reaching $7, billion. Increased exports of the processing and food industries are directly responsible for the overall increase of hard currency exports.

In 1990, the Soviet Union continued to be Hungary's chief trading partner, supplying mostly energy, raw materials and semi-finished products. The Soviet Union accounted for 19.9 percent of total trade in 1990. As of January 1, 1991, trade between Hungary and its CMEA trade partners is being transacted on a dollar-denominated basis.

During 1991, trade with the Soviet Union and other former-CMEA bloc member has nearly stopped. Exports to the West have soared, underscoring the increasing importance of the West to the Hungarian economy and the declining importance of CMEA. Hungary's ability to sustain this export momentum to the West could mitigate the falloff in exports to CMEA (and the weakening domestic demand component), halting the economy's downward trend.

C. Commercial Factors

U.S. Commercial Relations: In 1978, the United States and Hungary signed a trade agreement which provides for the reciprocal extension of non-discriminatory, most-favored-nation (MFN) treatment and contains provisions relating to trade expansion, business facilitation, industrial property and copyright protection, and arbitration of commercial disputes. Follow-up agreements, in 1979, deal with taxation and the formation of a Joint Economic and Commercial Committee (JECC) to facilitate the development of trade and economic cooperation.

The movement towards closer commercial ties has intensified. In October 1988, the United States signed a Memorandum of Understanding on Tourism with Hungary- to increase cooperation for the promotion of tourism. On November 28, 1989, the President signed into law the Support for East European Democracy Act (SEED). SEED channelled more than $900 million in support to Hungary and Poland for: structural adjustment; private sector development; trade and investment; educational, cultural and scientific activities; and other assistance programs. Within the context of SEED, the United States extended benefits of the Generalized System of Preferences (GSP) program to Hungary. Under this program, eligible products enter the United States duty-free from Hungary. In 1990, nearly $150 million worth of Hungarian products entered the United States duty-free. As a result of the 1991 GSP Annual Review, an additional $39 million in trade is eligible for GSP treatment. Additionally, SEED authorized Hungary's eligibility for Overseas Private Investment Corporation (OPIC) financing and insurance coverage. Finally, following a lengthy hiatus, the United States and Hungary will meet under the auspices of the JECC in late Spring/ Summer 1991.

This new direction of U.S.-Hungarian commercial relations has galvanized American business interest and activity. In 1989, U.S. exports to Hungary jumped 57.4 percent—the largest export growth in all of Eastern Europe. In 1990, U.S. exports increased 28.8 percent. For the first quarter 1991, U.S. statistics show that exports to Hungary climbed 85.8 percent. Since 1988, U.S. companies are participating, at an unprecedented rate, in joint ventures: Levi Strauss, Kodak, Schwinn Bicycle, Guardian Industries, Ford, GM, GE McDonald's, United Technologies, U.S. West, etc. The United States is the largest foreign investor in Hungary.

Hungary and The European Community: Not surprisingly, the re-orientation of the Hungarian economy from East to West is generating new economic ties with the European Community. In 1990. the proportion of products imported from the EC expanded from 29 to 33 percent of total Hungarian imports. Last year, Hungarian exports to the EC grew 26 percent in comparison to total hard currency exports which increased by 18 percent. This increase in exports is directly related to the modification of EC-Hungarian trade relations. In 1990, the EC provided Hungary with a number of concessions: lifting of discriminatory barriers; suspension of non-discriminatory barriers; extension of GSP; and, lifting of all self-restraint quotas.

During 1991, Hungary's relationship with tho Community continues to evolve. The EC and Hungary have undertaken negotiations to conclude an Association Agreement which will enter into force on January 1, 1992. A focal point of the talks is the establishment of an asymmetrical timetable (over an 8-10 year period) to eliminate tariffs between the EC and Hungary. The EC negotiators have resisted Hungary's calls to increase access to the EC agricultural and textile markets. The Association Agreement also addresses: cultural, economic and political cooperation; movements of capital, goods, services and people; and,, establishment of the institutional framework to manage the new relationship. The Agreement will lay the foundation for Hungary's eventual full membership in the Community.

D. Social/Cultural Factors

Hungary is a land-locked country in Central Europe, measuring 93,000 square kilometers. The capital city, Budapest (phonetics: BU-daw-pesht), resulted from the amalgamation of three cities in 1872: Buda, Obuda and Pest. Budapest is cut in half by the Danube ("Duna") river. Budapest not only serves as the seat of the national government, but is the cultural and economic center as well. Other sizeable Hungarian cities are: Debrecen, Miskolc, Pecs, and Szeged.

Hungary's population growth rate has been static for the last decade. About 5 million ethnic Hungarians live in neighboring countries (Czech and Slovak Republics, Romania, the Soviet Union and Yugoslavia) and the United States and Canada. Hungarians are considered to be among the most pessimistic people in the world—even when conditions are favorable. Consequently, American business representatives should not be frightened away by dire predictions. Hungarians are considered to be highly educated and technically proficient. Hungary is the home of the famous "rubic cube." Hungarian students have scored behind only Japan and Korea in mathematics.

The Hungarian language is not a European language but rather distantly related to finnish, estonian and mongolian. As a result of foreign occupation, some turkish and german derivations have crept into the language. In Hungarian, individuals are addressed by the last name followed by the first name (e.g., Mr. Smith, Joe). Hungarian business cards are presented in this manner unless printed in English.

IV. FINANCING

Hungary has accumulated a substantial hard currency debt. Recent increases in debt resulted from current account imbalances and exchange rate fluctuations. In May 1990, agreement was reached with the IMF (International Monetary Fund) on a standby loan which would help Hungary continue servicing its debt. In February 1991, Hungary signed a 3-year standby credit agreement with the IMF. The conditions of the agreement require that Hungary's current account and domestic deficits do not exceed $1.2 billion and HUF 78 billion, respectively.

Hungary is a member of the International Bank for Reconstruction and Development ("World Bank") and, as of fiscal year 1989, received loans totaling $345 million: During the past two years, Hungary has received additional loans from the World Bank. These loans are being used for project financing in the energy, food industry, other industrial sectors and on infrastructure (telecom, transport). In March 1990, a World Bank Structural Adjustment loan ($200 million) was made available to Hungary to support its stabilization program. The International Finance Corporation (IFC) also has been active in Hungary assisting primarily in industrial joint ventures and the financial sector. The Multilateral Investment Guarantee Agency (MIGA) provides political risk insurance for joint venture companies. Finally, the European Bank for Reconstruction and Development (EBRD) began operations in April 1991. The Bank's focus will be on stimulating private enterprise in Eastern Europe. The EBRD is an international Bank whose donors consist of the United States, Japan and the EC among other countries.

A two-tier banking system was reestablished in 1987. Until 1987, all banking activities were monopolized by the National Bank of Hungary. Comprehensive banking reform removed credit and service from the functions of the Central Bank. Enterprises can maintain commercial accounts with more than one bank, except for the ceilings on household deposits, commercial banks are free to set terms for deposits and loans. While companies with foreign participation may qualify for domestic currency loans from Hungarian banks (not subject to government approval), credit is in short supply and only on a short-term basis. The monopoly position of the National Bank in terms of foreign currency management and foreign lending is changing slowly. Foreign currency loans must be approved by the National Bank. West European banks are slowly entering the Hungarian market while, for the most part, U.S. banks are not yet active.

On June 22, 1990, the doors of the Budapest Stock Exchange were reopened. Having been closed since 1948, trade in securities was halted until 1983 when it was revived on a limited basis. Since 1987, securities have played an increasingly important role in supplementing capital for companies. New ventures into this market include: Creditanstalt Bankverein's (Austria) joint venture with Budapest Bank—CA-BB, Ltd; LUPIS BROKERHAZ; and a new Hungarian company called Co-Nexus. The Exchange's volume and listed shares, however, remain small.

Other sources of financing for U.S. ventures in Hungary include:

Hungarian-American Enterprise Fund (HAEF)—established to support the SEED Act for the purpose of promoting the development of private enterprise in Hungary, including small- and medium-size business, the agricultural sector and U.S.-Hungarian joint ventures.

Overseas Private Investment Corporation (OPIC)—key programs are its loan guarantees, direct loans and political risk insurance. The Eastern European Growth Fund is designed to OPIC funds with private venture capital to finance new business enterprises.

Trade and Development Program (TDP) - an independent U.S. Government agency assists firms in identifying major development programs which offer export potential and by funding U.S. private sector involvement in planning.

V. TRADE AND INVESTMENT ISSUES/BARRIERS

Key elements of Hungary's economic program are the liberalization of the trade and investment regimes.

On December 23, 1990, the Government issued Decree Number 112, continuing its liberalization program. The list of products subject to import licensing procedures has been further reduced. Currently, 93 percent of imported products can be imported without prior approval. Since 1987, the IMPEXs' (government-sanctioned foreign trade organizations) monopoly of foreign trade has gradually been abolished. Decree Number 112 maintains exclusive trading rights for the following product areas: primary energies and fuels (e.g., black/brown coke coal); precious metal jewelry and precious stones; products restricted on the basis of environmental or health considerations (e.g., cyanide, aliphatic compounds); and products restricted pursuant to international agreement (e.g., pork to the EC only). For a complete listing of products subject to import licensing, monopoly trade and export restrictions, U.S. businesses are urged to contact EEBIC or the U.S. Embassy.

Hungary is very receptive to foreign investment. Hungary is incorporating Western practices and business safeguards into its legal code. Consequently, issues traditionally addressed in a bilateral investment treaty, BIT, (e.g., expropriation, repatriation of profits, etc.,) to gain market access and to safeguard U.S. business interests, are not problematical in Hungary.

The U.S. Government continues to monitor developments to ensure that actions by the Government of Hungary do not adversely impact or violate U.S. trade and investment rights.

A. Trade Barriers

(1) Import Licensing—Hungary requires import licenses for certain imported products. Licenses are granted on a first-come/first-serve basis. As of January 1991, Hungary has eliminated licensing requirements on over 93 percent of imports.

(2) Tariffs—Hungary is a contracting party to the General Agreement on Tariffs and Trade (GATT). Hungary has bound approximately 87 percent of its tariff schedule—in other words, committed not to raise tariffs about these bound fates. Hungary's maximum tariff rates fall in the 40-50 percent range, although the majority of Hungary's tariffs are under 20 percent, averaging 18 percent. The Government recently announced that Hungary would maintain its GATT obligations and reduce the average tariff to around 8 percent. However, following this announcement, the Government increased tariffs on automobiles from 10 to 18 percent (13 percent on environmentally-friendly autos). The Government also notified the GATT that it intended to raise tariffs on 4 product categories—3 of these tariff increases would violate Hungary's current GATT commitments.

The United States has requested a reduction of Hungary's tariffs in the Uruguay Round of multilateral trade negotiations.

(3) Convertible currency trade transactions must be registered with the Ministry of International Economic Relations (MIER). Registrants must provide the following information on planned trade activities: products for import or export: proposed trading partners; and, projected annual volumes. Firms must also attest to their ability (organizational structure, personnel qualifications) to engage in foreign trade. For liberalized products, the registration process is virtually automatic. MIER's permission is also required for trading products subject to government control.

(4) Lack of Adequate IPR Protection—Hungary provides process patent protection but not product patent protection. This has been particularly problematic in the pharmaceuticals sector. Hungary is drafting a new patent law but enactment could be delayed depending on other Parliamentary business and pressure from the politically powerful pharmaceutical industry.

The U.S. government has engaged Hungary in bilateral discussions on IPR issues.

(5) U.S. businesses have complained that customs officials' ignorance of regulations and lack of convenient customs facilities sometimes hinder business activities.

(6) Potential Quotas on Consumer Goods—Section 7 of Decree 112 states that MIER may impose a half yearly limit on the importation of consumer goods. Furthermore, section 7 empowers MIER to license additional quotas in "cases which are justified."

The U.S. government will monitor the situation, if this law is utilized by the GOH, to ensure that U.S. rights are not violated.

B. Investment Issues

(1) Convertibility—The Hungarian Forint is not a freely convertible currency. However, the Government has implemented provisions for companies to repatriate hard currency profits. Hard currency imports by a joint venture are subject to prior approval from MIER.

(2) Infrastructure—The poor state of the transport and telecommunications infrastructures, as well as the housing and general real estate situation, act as the most effective barriers to U.S. investment. Recognizing that these weaknesses undermine Hungary's endeavors to attract foreign investment, the Government is undertaking costly improvements.

VI. MARKET ANALYSIS PLAN (MAP)

A. Industry Sector Analyses (ISAs)

1. Title: Paper & Paperboard
Code: PAP
Completion Date: December 1991
Country: Hungary
Researcher: Andrea Imrik
Work days needed for research: 40
Justification: The U.S. paper/paperboard industry is highly competitive. Liberalization of the market and the uncompetitiveness of the Hungarian industry have combined to create opportunities for U.S. exporters.

2. Title: Computer Software & Services
Code: CSF
Completion Date: March 1992
Country: Hungary
Researcher: Agnes Dalos
Work days needed for research: 45
Justification: Information technology is outdated. New opportunities in the hardware segment of the market will be accompanied by increased demand for software. U.S. software producers continue to be the most seminal and competitive internationally.

3. Title: Agricultural Equipment
Code: AGM
Completion Date: April 1992

Country: Hungary
Researcher: Kinga Svastics
Work days needed for research: 40
Justification: There is renewed emphasis on increasing the productivity of the agricultural sector which—the primary hard currency earner in the economy. In 1990, increased agricultural exports largely accounted for Hungary's trade surplus. The GOH continues to target this sector.

4. Title: General Consumer Goods
Code: CGC
Completion Date: February 1992
Country: Hungary
Researcher: Not Determined
Work days needed for research: 40
Justification: Market opportunities will rapidly expand in the future given the current pent-up demand

5. Title: Plastic Production Machinery
Code: PME
Completion Date: July 1992
Country: Hungary
Researcher: Not Determined
Work days needed for research: 45
Justification: Plastics as a by-product of the chemical industry which is very competitive in Hungary. Plastics manufacturers require more technologically sophisticated equipment.

6. Title: Dairy Processing Equipment
Code: FPP
Completion Date: September 1992
Country: Hungary
Researcher: Not Determined
Work days needed for research: 40
Justification: Subsector of Best Prospects.

7. Title: Building Products
Code: BLD
Completion Date: June 1992
Country: Hungary
Researcher: Katalin Barazda
Work days needed for research: 45
Justification: Related sector to Best Prospects "Construction Equipment."

VI. TRADE EVENT PLAN

1. Event: Print '91
Event Dates: September 4-11, 1991
Industry Theme: Printing
Industry Code: PGA
Event Type: FBP

Location: Chicago, Illinois
Washington Recruiter: N/A
Export Background of Participants: N/A
Estimated Workdays: 14
Estimated Budget: N/A

2. Event: USA Showcase '91
Event Dates: October 17-21, 1991
Industry Theme: Computers, Telecom, Franchising, Medical, Pharmaceutical, Environmental and Electronic.
Industry Code: Z
Event Type: Trade Fair
Location: Budapest
Washington Recruiter: Frederica Wheeler
Export Background of Participants: N/A
Estimated Workdays: 3
Estimated Budget: N/A

3. Event: International Winter Consumer Electronics Show
Event Dates: January 9-12, 1992
Industry Theme: Consumer electronics equipment/services
Industry Code: CEL
Event Type: FBP
Location: Las Vegas, Nevada
Recruiter: Agnes Dalos (USFCS/Budapest)
Export Background of Participants: N/A
Estimated Workdays: 14
Estimated Budget: N/A

4. Event: Communication Network '92
Event Dates: January 28-30, 1992
Industry Theme: Communication equipment/services
Industry Code: TEL
Event Type: FBP
Location: Washington, DC
Recruiter: Agnes Dalos (USFCS/Budapest)
Export Background of Participants: N/A
Estimated Workdays: 14
Estimated Budget: N/A

5. Event: International Fluid Power Exhibition '92
Event Dates: March 24-26, 1992
Industry Theme: Hydraulic equipment
Industry Code: Z
Event Type: FBP
Location: Chicago, Illinois
Recruiter: Andrea Imrik (USFCS/Budapest)
Export Background of Participants: N/A
Estimated Workdays: 14
Estimated Budget: N/A

6. Event: Capital Goods Solo Exhibition
   Event Dates: Spring 1992
   Industry Theme: Capital Goods
   Industry Code: Z
   Event Type: Exhibition
   Location: Bratislava
   Washington Recruiter: Joseph Harrison, 377-2474
   Export Background of Participants: N/A

7. Event: IFABO
   Event Dates: April 22-25, 1992
   Industry Theme: Computers and Office Equipment
   Industry Code: CPT
   Event Type: Trade Fair
   Location: Budapest
   Washington Recruiter: IDG World Expo
   Export Background of Participants: N/A
   Estimated Workdays: 35
   Estimated Budget: $70,000

8. Event: Waste Expo '92
   Event Dates: May 5-8, 1992
   Industry Theme: Waste handling equipment services
   Industry Code: POL
   Event Type: FBP
   Location: New Orleans, Louisiana
   Recruiter: Kinga Svastics (USFCS/Budapest)
   Export Background of Participants: N/A
   Estimated Workdays: 14 Estimated Budget: N/A

9. Event: Budapest International Spring Fair
   Event Dates: May 20-28, 1992
   Industry Theme: Capital Equipment
   Industry Code: Z
   Event Type: Trade Fair
   Location: Budapest
   Washington Recruiter: Russell Johnson, 377-2159
   Export Background of Participants: N/A
   Estimated Workdays: 40
   Estimated Budget: $200,000

10. Event: State of Ohio Trade Mission
    Event Dates: May 23-25, 1992
    Industry Theme: Various
    Industry Code: Z
    Event Type: SOGA
    Location: Budapest
    Recruiter: State of Ohio
    Export Background of Participants: N/A
    Estimated Workdays: 12
    Estimated Budget: $4,500

11. Event: The Danube Basin Conference
    Event Dates: June, 1992
    Industry Theme: Various
    Industry Code: Z
    Event Type: Conference
    Location: Vienna, Austria
    Washington Recruiter: EEBIC
    Export Background of Participants: N/A
    Estimated Workdays: 12
    Estimated Budget: N/A

12. Event: Semicon West '92
    Event Dates: June 16-18, 1992
    Industry Theme: Semiconductor manufacturing, testing & assembly equip
    Industry Code: ELC
    Event Type: FBP
    Location: San Francisco, California
    Recruiter: Agnes Dalos (USFCS/Budapest)
    Export Background of Participants: N/A
    Estimated Workdays: 14
    Estimated Budget: N/A

13. Event: Consumer Goods Trade Mission
    Event Dates: Summer 1992 (not confirmed)
    Industry Theme: Consumer Goods
    Industry Code: GCG
    Event Type: Trade Mission
    Location: Budapest
    Washington Recruiter: Office of Consumer Goods, 377-1176
    Export Background of Participants: N/A
    Estimated Workdays: 20
    Estimated Budget: N/A

14. Event: IMTS '92
    Event Dates: September 9-17, 1992
    Industry Theme: Machine tools
    Industry Code: MTL
    Event Type: FBP
    Location: Chicago, Illinois
    Recruiter: Andrea Imrik (USFCS/Budapest)
    Export Background of Participants: N/A
    Estimated Workdays: 14
    Estimated Budget: N/A

VII. THE WORK PLAN

Critical Element I: Market analysis and program planning. Analyze host-country environment and plan appropriate marketing strategy. (10 Percent)

- Core program
  1. Develop the Country Marketing Plan and revise as necessary during the year.
  2. Report host-country trade barriers and trade and investment policies.
  3. Assess the commercial and economic environment and provide host-country trade data.

- Country-specific activities for current year
  1. Track economic reforms and report on new laws and regulations affecting trade and investment with particular focus on: amendments to the Investment Act (how joint ventures will be taxed); how privatization is implemented by SPA; new laws on land ownership, etc.,
  2. Report on competitors' activities and development of host-country's international commercial relations with the EC and East Asian countries.
  3. Report on host-country's industrial restructuring plans and activities, to include U.S. companies' involvement- and success in bidding on GOH issued tenders.
  4. Contribute to Embassy's general reporting on economic and commercial matters.

Critical Element II: U.S. Government commercial program support. Develop and manage trade and investor policy programs and activities. (25 Percent)

- Core Program
  1. Provide program and visitor support to U.S. government agencies, including IEP, TD, IA, TDP, EximBank, OPIC, NOAA, PTO, USTTA, USTR, NTIA, NIST, and BXA.
  2. Carry out the overseas export-control responsibilities of the Department as identified in the ITA-BXA memorandum of understanding.
  3. Support and participate in bilateral and multilateral trade negotiations.
  4. Report on and support major project opportunities.
  5. Provide support to host-country government official visits to the U.S.

- Country-specific activities
  1. Support and participate in next meeting of the Hungarian JECC.
  2. Support and participate in negotiations between the U.S. and Hungary to conclude a comprehensive trade and investment agreement which addresses services and intellectual property rights.
  3. Report on export licensing problems U.S. firms are experiencing and assist BXA with country-specific foreign availability investigations.

Critical Element III: Business counseling and outreach. Counsel and assist companies in increasing U.S. exports of goods and services and in achieving their international business interests. (30 Percent)

- Core Program
  1. Maintain key host-country government and business contacts.
  2. Counsel and assist U.S. business visitors.
  3. Advocate on behalf of U.S. business interests.
  4. Support business activities related to licensing, joint ventures, and export-related investment.
  5. Assist in resolving commercial trade complaints.
  6. Serve as liaison with the AmCham and other local business groups.

- Country-specific activities
  1. Support and participate in the activities of the American Chamber of Commerce.

2. Provide assistance and counsel U.S. business representatives establishing offices and taking-up residence in Host country.
3. Provide introductions to high-level government and business contacts.
4. Assist in scheduling of business meetings for U.S. company representatives.
5. Assist U.S. firms in obtaining World Bank tenders and follow up on bids submitted.
6. Introduce "Gold Key" service.

Critical Element IV: Trade promotion and commercial information support. Initiate, plan and manage U.S. trade promotion and commercial information programs, with emphasis on infrequent exporters. (20 Percent)

- Core program
  1. Provide commercial information and contact services, to include ADSs and WTDRs.
  2. Develop and carry-out a market research program (ISAs, market insight information, alert reports, CSS, etc.).
  3. Identify and report trade opportunities under the revised TOP.
  4. Disseminate the information in "Commercial News USA."
  5 Develop and carry-out market-relevant.trade promotion events.

- Country-specific activities
  1. Recruit for Budapest Spring Fair, IFABO and Showcase '91.
  2. Undertake seven ISAs and supervise the course of the project.
  3. Continue with introduction of ADSs, CSSs, and WTDRs to Post.
  4. Assist in planning and staging individual company trade promotion events, meetings, and conferences.
  5. Support and participate in EE regional trade promotion activities and meetings.

Critical Element V: Management, supervision and administration. Manage and administer the post's commercial program. (15 Percent)

- Core program
  1. Manage program and personnel resources.
  2. Ensure that the Department's policies and requirements regarding equal employment opportunities are carried out at pos t.
  3. Manage administrative and budget requirements.
  4. Submit quarterly statistical reports, and scheduled reports on event receipts, expenditures and event scheduling changes.
  5. Report on accomplishments (monthly highlights, success stories, end-of-show reports).
  6. Prepare performance plans and conduct midyear progress reviews and annual performance appraisals.
  7. Maintain and manage the Commercial Information Management System (CIMS).
  8. Plan for and provide training opportunities.
  9. Participate in and support country-team activities.

APPENDIX L - MARKETING PLAN: CZECHOSLOVAKIA

This appendix was made available by the U.S. and Foreign Commercial Service, Prague, Czechoslovakia. It is dated November 1991.

I. COUNTRY DATA

A. Profile

Country Name:	Czech and Slovak Federal Republic (CSFR)
Population:	15.7 Million
Religion:	50 % Roman Catholic; 20 % Protestant
Government:	Transitional Parliamentary Democracy
Language:	Czech, Slovak
Work Week:	Monday to Friday
Contacts:	
SCO:	Robert M. Shipley Phone: 011-42-2-536-641/9; after hours 536-646 Fax: 011-42-2-532-457; or 537-534 Location: U.S. Embassy
PCO:	Shelley Galbraith
IEP Desk Officer:	Mark Mowrey Phone: 202-377-4915 Fax: 202-377-2155; 202-377-4098 Location: U.S. Department of Commerce, Washington, DC
FCS Regional Mgr.:	Charles Ford
Director Non-EC Europe:	George Knowles Phone: 202-377-1599 Fax: 202-377-3159 Location: U.S. Department of Commerce, Washington, DC.

B. Domestic Economy	88	89	90	91(E)
GDP (Billion Crowns-IMF)	598.6	609.6	798.1	619.25
GDP Growth Rate (%)	1.4	1.0	30	-14
GNP Per Capita (IMF)	10,130	7,878	8,200	N/A
Govt. Spending as % of GDP	N/A			
Inflation (%)	0.3	1.5	12	60
Unemployment	0	0	2	8
Foreign Exchange Reserve (Mill. SDR)		1.4		
Average Exchange Rate (KCS/USD):		14.9	17	30.5
Foreign Debt (Bil. DLS.)	7.2	7.8	8.1	11
Debt Service Ratio	24	29	32	25
U.S. Economic Assistance (Millions USD):				150 (E)
U.S. Military Assistance: N/A				

(E) Estimated

C. Trade

	88	89	90	91(E)
Total Exports (Bil. DLS.)	14.7	14.6	12.7	N/A
Total Imports (Bil. DLS.)	14.4	14.4	14	
Exports to U.S. (CIF) (Million DLS.)	87.6	87	87	
U.S. Share of CSFR Imports (%)	.003	.003	.01	

Imports of Mfg. Goods: N/A
Leading Trading Partners: USSR, Germany, Poland
Principal U.S. Exports: Cattle Hides, Fertilizers, Cotton, Tobacco, Analytical Equipment
Principal U.S. Imports: Glassware, Footwear, Tractors, Tires, Apparel

Note: The CSFR received most favored nation status (MFN) for its exports to the U.S. in November, 1990. It also received duty-free or reduced tariff status for the export of approximately 400 products under the U.S. generalized system of preferences (GSP) in April 1991.

D. Investment

Total Foreign Direct Investment:	Approximately USD 3 Billion
U.S. Direct Investment:	Approximately 400 Million Dollars*
Principal Foreign Investors:	Germany, Austria, Switzerland, United States

*U.S. Investment is committed, not actual, capital. Largest investors are: DOW Chemical, 200 Million USD; USWEST/Bell Atlantic, 80 Million USD, Procter and Gamble, 44 Million USD; and Ralston Energy Systems (Eveready), 24 Million USD.

Sources: IMF, Planecon, U.S. Dept. State, Czech Republic Ministry of Economic Policy; Federal Ministry of Foreign Trade.

II. BEST PROSPECTS

These product and service categories are ranked in priority order. No statistics exist on market size and growth. All are new markets for U.S. exporters, and all should increase dramatically in size over the next few years. On a scale of 5 (Highest) to 0 (Lowest), the CSFR's receptivity to all these products/services is 5. No hard data exists to indicate size of market and potential. Also, these markets are not traditional markets for Western products, and represent new areas for U.S. Sales based on dramatic economic change in the country.

1. Computers and Periipherals (CPT)

A. Rank of Sector: 1
B. Name of Sector: Computers and Peripherals
C. Code: CPT
E. Est. Average Growth Rate of Market: 30%
J-1. Receptivity: 5
J-2. Competition: 1
J-3. Market Barriers: 5
K. Comments: The need for computers to upgrade nearly all businesses, government functions, and factories means that computers and applications software are in very high demand. Particularly in demand are low-cost alternatives to high MIS costs. This traditional U.S. export now faces low export control barriers and high brand recognition in local market.

L. Most Promising Subsectors:
   1. Mainframes
   2. Microcomputer systems
   3. Minicomputer systems
   4. Personal Computers
   5. Computer Peripherals
   6. Computer Printers

2. Pollution Control Equipment
A. Rank of Sector: 2
B. Name of Sector: Pollution Control Equipment
C. Code: POL
J-1. Receptivity: 5
J-2. Competition: 2
J-3. Market Barriers: 3
K. Comments: Government priorities have been set to reduce serious air pollution, water pollution and solid waste disposal problems and to remediate existing waste sites.
L. Most Promising Subsectors:
   1. Air Pollution Control Equipment
   2. Water Pollution Control Equipment
   3. Toxic Waste Treatment Materials/Equipment
   4. Water Purification Equipment
   5. Solid/Hazardous Waste Management
   6. Industrial Waste Treatment Equipment

3. Computer Software and Services (CSF)
A. Rank of Sector: 3
B. Name of Sector: Computer Software and Services
C. Code: CSF
J-1. Receptivity: 5
J-2. Competition: 1
J-3. Market Barriers: 5
K. Comments: Same as Computers and Peripherals (CPT)
L. Most Promising Subsectors:
   1. Applications Software
   2. Business Applications Software
   3. Financial Software
   4. Data Processing Management and Maintenance
   5. Programming Utility Software

4. Telecommunications Equipment (CPT)
A. Rank of Sector: 4
B. Name of Sector:Telecommunications Equipment
C. Code:TEL
J 1. Receptivity:5
   2. Competition:1
   3. Market Barriers:5
K. Comments: CSFR's telecommunications industry needs complete overhaul. Bell Atlantic/U.S. West are implementing cellular and packet switching systems. Hardware is needed for implementation. New broadcasting law clarifies regulation of the airwaves and private stations. Satellite communications also scheduled for modernization. Fiber optics system will be an area for future focus.

L. Most Promising Subsectors:
   1. Cellular Telephone Equipment
   2. Telecom. Network Switching/Transmission Equipment
   3. Telephone Switching Equipment
   4. Fiber Optic Transmission Equipment
   5. Telephone Equipment
   6. Telephone/Teleprinter/Fax Equipment
   7. Broadcast Equipment
   8. Satellite Telecommunications Equipment

5. Telecommunications Services (TES)
A. Rank of Sector:5
B. Name of Sector:Telecommunications Services
C. Code:TES
J 1. Receptivity:5
   2. Competition:1
   3. Market Barriers:5
K. Comments: Same as for telecommunications equipment
L. Most Promising Subsectors:
   1. Mobile Communications Service
   2. Value-added network services

6. Electrical Power Systems (ELP)
A. Rank of Sector:6
B. Name of Sector:Electrical Power Systems
C. Code:ELP
J 1. Receptivity:5
   2. Competition:1
   3. Market Barriers:3
K. Comments: CSFR seeks to expand its nuclear generating capability, modernize its traditional coal-fired generators, and install gas-driven equipment.
L. Most Promising Subsectors:
   1. Thermal Power Projects
   2. Electric Generating Equipment
   3. Electric Distribution and Transmission Equipment

7. Information Services (INF)
A. Rank of Sector:7
B. Name of Sector:Information Services
C. Code:INF
J 1. Receptivity:5
   2. Competition:1
   3. Market Barriers:5
K. Comments: Government and emerging private sector are literally desperate for Western data bases.
L. Most Promising Subsectors:
   1. On-Line Data Bases

8. Education and Manpower Training Services (EDS)
A. Rank of Sector:8
B. Name of Sector:Education and Manpower Training Svc.

C. Code:EDS
J 1. Receptivity:5
2. Competition:2
3. Market Barriers:5
K. Comments: All sectors of society need education and training in market economy and free enterprise, especially with the on-going privatization of state-owned companies.

9. Management Consulting Services (MCS)
A. Rank of Sector:9
B. Name of Sector:Management Consulting Services
C. Code:MCS
J 1. Receptivity:5
2. Competition:2
3. Market Barriers:5
K. Comments: As private firms emerge, all need assistance in switching to market and profit orientation; accounting for performance; managing/training employees; and coping with competition, supply and demand.
L. Most Promising Subsectors:
1. Financing/Accounting Management Services
2. Plant Management
3. Privatization

10. Financial Services (FNS)
A. Rank of Sector:10
B. Name of Sector:Financial Services
C. Code:FNS
J 1. Receptivity:5
2. Competition:2
3. Market Barriers:4
K. Comments: The banking sector is in transition. State-owned banks will be privatized, many in partnership with foreign banks. Foreign banks are being licensed to operate. Absence of hard currency will restrict commercial lending. Training of bankers, especially in management and international trade is primary.

11. Travel and Tourism Serivces (TRA)
A. Rank of Sector:11
B. Name of Sector:Travel and Tourism Services
C. Code:TRA
J 1. Receptivity:5
2. Competition:1
3. Market Barriers:4
K. Comments: More than 50 million tourists are estimated in 1991. Prague has an estimated shortage of 12,000 hotel beds per day. Hotels and inns are scarce to nonexistant in smaller cities and towns. The country's natural beauty, history and culture make it a tourist's paradise. Stiff competition in hotel construction/management from Austria, France and Germany.

12. Aircraft and Parts (AIR)
A. Rank of Sector:12
B. Name of Sector:Aircraft and Parts
C. Code:AIR
J 1. Receptivity:5

2. Competition:1
3. Market Barriers:4

K. Comments: CSA (National Airline) will be privatized. New medium-range aircraft will be leased. New private airline (Tatra Air) now operating in Moravia and Slovakia, operating small commuter planes. Relatively sophisticated domestic aircraft industry, manufacturing airplanes, engines, avionics. U.S. firms already involved in engine manufacture, avionics supply, parts manufacture. Helicopters also needed. Good market for military sales also.

L. Most Promising Subsectors:
1. Commercial Passenger/Cargo Jet Aircraft
2. Executive/General Aviation Aircraft
3. Aircraft Jet Engines
4. Helicopters
5. General Aviation

13. Avionics and Ground Support Equipment (AVG)

A. Rank of Sector: 13

B. Name of Sector: Avionics and Ground Support Equip.

C. Code: AVG

J 1. Receptivity: 5
2. Competition: 1
3. Market Barriers: 4

K. Comments: Dramatic traffic increase means need for new air traffic control systems, larger terminals, longer runways, new airports, more ground support, and better maintenance facilities.

L. Most Promising Subsectors:
1. Air Traffic Control/Navaids
2. Avionics
3. Airport Control Systems
4. Airport Ground Services Equipment
5. Aviation Ground Support Equipment
6. Commercial Aviation Maintenance

14. Medical Equipment (MED)

A. Rank of Sector: 14

B. Name of Sector: Medical Equipment

C. Code: MED

J 1. Receptivity: 5
2. Competition: 1
3. Market Barriers: 5

K. Comments: Health care system will be privatized, but more slowly than industry. Some form of socialized medicine will remain. Need for equipment is great, but little means to pay, especially in rural areas and smaller cities.

L. Most Promising Subsectors:
1. Cardiological/Cardiovascular Equipment
2. Diagnostic Equipment
3. Anesthesia and Resuscitation Equipment
4. Patient Monitoring Systems/Apparatus
5. Surgical Equipment
6. Rehabilitations Devices and Equipment

15. Franchising (FRA)
A. Rank of Sector: 15
B. Name of Sector: Franchising
C. Code: FRA
J 1. Receptivity: 5
   2. Competition: 2
   3. Market Barriers: 4
K. Comments: Growth in tourism and local interest in and demand for Western products creates limitless opportunities for restaurants, leisure clubs, and specialty stores. Local disposable income is still low (avg. salary 150-200 USD per month), but Western tourist traffic is overwhelming. Good fast food restaurants are few. Local food raw materials are good and plentiful. Health clubs are virtually nonexistant outside of major hotels.
L. Most Promising Subsectors:
   1. Fast Food Franchises
   2. Clothing/Specialty Products
   3. Health Clubs

16. Accounting Services (ACT)
A. Rank of Sector: 16
B. Name of Sector: Accounting Services
C. Code: ACT
J 1. Receptivity: 5
   2. Competition: 3
   3. Market Barriers: 5
K. Comments: Big six accounting firms are active in privatization counseling. Western accounting methods are virtually unknown. The thousands of emerging private enterprises will need to learn them.

17. Drugs and Pharmaceuticals (DRG)
A. Rank of Sector: 17
B. Name of Sector: Drugs and Pharmaceuticals
C. Code: DRG
J 1. Receptivity: 5
   2. Competition: 1
   3. Market Barriers: 3
K. Comments: See medical equipment (MED). Import licenses required. Local manufacture in joint-venture possible, but concern exists for copyright/patent protection under current laws. Heavy competition from Germany and Switzerland.
L. Most Promising Subsectors:
   1. Antibiotics
   2. Supplies

18. Food Processing and Packaging Equipment (FPP)
A. Rank of Sector: 18
B. Name of Sector: Food Processing and Packaging Equip.
C. Code: FPP
J 1. Receptivity: 5
   2. Competition: 1
   3. Market Barriers: 5
K. Comments: Great demand for equipment to develop a local industry which has excellent potential for export, especially meats, dairy products, fruits/vegetables, beer, and wine. Heavy competition from European community suppliers, mainly Italy.

L. Most Promising Subsectors:
   1. Meat and Poultry Processing Equipment
   2. Beverages FPP Machinery
   3. Bottling and Container Packaging Equipment
   4. Fruit-Vegetable Processing Equipment
   5. Dairy Processing Equipment
   6. Food Processing Plant Storage Facilities

19. Hotel and Restaurant Equipment (HTL)
A. Rank of Sector: 19
B. Name of Sector: Hotel and Restaurant Equipment
C. Code: HTL
J 1. Receptivity: 5
   2. Competition: 1
   3. Market Barriers: 5
K. Comments: See travel and tourism services (TRA) and franchising (FRA). Stiff competition from European community suppliers.
L. Most Promising Subsectors:
   1. Restaurant Food Preparation/Service Equipment
   2. Food Preparation Equipment: Hotel/Restaurant
   3. Hotel Automation
   4. Fast Food Equipment
   5. Hotel and Institutional Catering Equipment

20. Renewable Energy Equipment (REQ)
A. Rank of Sector: 20
B. Name of Sector: Renewable Energy Equipment
C. Code: REQ
J 1. Receptivity: 5
   2. Competition: 3
   3. Market Barriers 5
K. Comments: Federal energy agency director mentions renewable energy generation as a priority area. See also electrical power systems (ELP).
L. Most Promising Subsectors:
   1. Biomass Energy Generation Equipment

III. COMMERCIAL ENVIRONMENT

A. Political Factors

Czechoslovakia has several ethnic groups, Bohemians, Moravians and Silesians mainly in the Czech lands, and Slovaks and other minorities in Slovakia. With the Democratic change to Czechoslovakia's political system has given rise to a political movement for the independence of Slovakia, one of the two republics in the country. The Slovak question remains open. The political pressure to recognize the desire of Nationalist Slovaks for Independence remains, and its effect on investment and foreign business relations for Czechoslovakia is unclear. Certainly today businesses do not see a greatly increased risk due to the possibility that the country might split apart. Public opinion polls show that a majority of citizens, both Czech and Slovak, favor the currently established form of Federal Republic Government, but there are those who favor a Confederation, or an independent Slovak state. The question is yet to be resolved, and the government is considering a popular referendum to put the question to rest once and for all.

The result of this is clear for business: companies must consider doing business in each republic, visiting both Prague and Bratislava, the capital of Slovakia, to explore opportunities in each part of the country. In addition, business representatives must be mindful of the ethnic composition of the country, referring at all times to its citizens as "Czechs and Slovaks" or "Czechoslovaks."

B. Economic Factors

The United States granted the CSFR provisional most favored nation status in November, 1990. With Congressional removal of the CSFR from the Jackson-Vanik legislative restrictions in the fall of 1991, the CSFR will have permanent MFN status.

U.S. investment promotion in the CSFR was boosted by the October, 1991 signing of the U.S./CSFR Bilateral Investment Treaty (BIT) and an agreement with the U.S. Overseas Private Investment Corporation (OPIC, see section IV.E.). Under the new commercial code taking effect in January, 1992, government approval of foreign investment will no longer be required, except for the direct sale of government equity. The CSFR government's official position is that all large domestic enterprises now being privatized (some 4,000 in both republics over a two year period representing 95 percent of the economy) are open to foreign investors as minority or majority investment partners or full owners. The number of U.S. companies with representative offices in Prague has grown from 6 at the time of the revolution in November 1989 to nearly 200 in November 1991.

C. Commercial Factors

Tax Reform: The tax system was revised recently to the benefit of joint-venture enterprises. The corporate income tax or profit tax, which is 55 percent for domestic companies, was reduced to 40 percent for companies with a minimum 30 percent foreign capital participation. New joint-venture companies may apply for a two year tax exemption on the profits tax, effective from the date the company is legally established in the CSFR. Savings based on this exemption cannot be taken out of the country, but must be reinvested. A profit tax of 50 percent exists for agricultural enterprises.

There is a uniform payroll tax rate of 50 percent. Four basic turnover tax rates exist (0, 11, 20, and 29 percent) depending on the product. The turnover tax is an annual tax on total sales volume, which must be reported to fiscal authorities by February 28 for the previous year.

There is a uniform withholding tax for profits taken out of the country of 25 percent. A uniform value added tax (VAT) will be implemented in 1993.

The income tax for individuals is progressive—15 percent on yearly profit up to 60,000 crowns (1,967 dollars), and increasing to a maximum of 55 percent on profits over 1,080,000 crowns (35,410 dollars).

The CSFR tax system will be revised over the next year, moving toward a uniform value added tax structure in 1992. The U.S. and CSFR governments are negotiating a bilateral tax treaty which should be signed during 1992. In addition to other provisions, this treaty will deal with the issue of double taxation for investors in each country.

Foreign trade organizations: Monopolies on foreign trade granted to these state-owned organizations have been abolished. Most of the 52 government trading companies are flourishing, however, as they are gradually transformed into joint-stock companies. Part of their shares will continue to be held by the state, but most will be purchased by banks and manufacturing companies who utilize their services.

Tariff structure/customs procedures/foreign trade environment: the CSFR government plans to restructure its tariff rates to protect sensitive industries as the economy moves toward a market-oriented base. A second motivation is to move away from the communist era system of controlling imports largely by limiting access to foreign exchange. The CSFR plans to apply for a Gatt waiver to increase its overall tariff base from 5 to 5.7 percent. While this increase appears nominal, there is substantial opposition by Gas members to a permanent increase in the CSFR's tariffs because of the large increases on products representing promising import markets. CSFR officials have given verbal assurance their government will participate fully in Gatt to lower tariff rates over the next decade as economic conditions permit.

In addition, the CSFR government amended an existing 15 percent, customs surcharge for imported consumer goods and foodstuffs sold in the CSFR. The surcharge was raised to 20 percent for commercial goods and was applied at 15 percent for private imports for personal use, with a duty-free limit of 100 dollars. These measures were implemented to conserve scarce foreign exchange reserves.

It will be several years before the crown may achieve full convertibility to Western currencies. This is chiefly due to the CSFR's inability to earn foreign exchange, the high cost of oil imports, the disappearance of subsidized oil from the USSR, the loss of East Germany as one of the CSFR's chief customers, and a significant reduction in exports to the USSR.

Access to foreign trading activities is greatly liberalized. As of February 1, 1991, every agent registered in the CSFR register of enterprises is entitled to engage in foreign trade without other registration or authorization, except for the following goods: weapons, drugs, alcohol, tobacco, and nuclear materials, where specific authorization is required from the Federal Ministry of Foreign Trade.

D. Social/Cultural Factors

Important guidance for U.S. Businesspersons Contemplating Operations in the CSFR: U.S. business representatives in the CSFR are traditionally at a disadvantage vis-a-vis their peers from Germany, Austria, and France. This situation shows little improvement despite a strong desire by senior CSFR government officials to increase trade with the United States.

We have discussed informally with senior CSFR government officials the anatomy of several recent business deals—some won by U.S. firms and some lost. Our conversations ignored specific criteria such as price, scope of work, terms of payment, etc., concentrating instead on attitudes, reactions, and relationships. Allowing for the subjective skew of personal opinion, the inaccuracy of generalities, and the peculiarity which applies to each negotiation, several consistent themes of criticism emerge. The following guidance is meant as constructive suggestion for any U.S. businessperson in the CSFR.

Firstly, initial business contacts should not be solely senior executive to senior management. The dynamics of commercial and industrial decision-making here are seldom energized from the top. Active support of middle management and, where relevant, labor is far more critical. As we see daily, top executive authority can easily be hamstrung from the bottom, and top management override of opinion emerging from this base is rare.

Secondly, regional sensitivity is regularly underestimated. Business deals in Slovakia, for example, usually require simultaneous and balanced negotiations with both Federal and Slovak government or industry representatives. Ignoring the necessity for this multiparty involvement is to guarantee failure.

Thirdly, there seems to exist at highest CSFR government and entreprenurial levels a pervasive sensitivity that U.S. business damages its chances by conveying a condescending attitude; by creating the illusion that it is doing the CSFR a favor by simply appearing in the country. Officials with whom we speak admit to relative inexperience in free market business. They, however, express deep resentment at being treated too often with what they regard as intellectual arrogance and commercial impatience. They also repeatedly assert that U.S. businesspeople do not negotiate, but attempt to dictate.

Fourthly, we often hear the complaint that U.S. businesspeople "arrive on Monday, have meetings on Tuesday, and expect to take their money out on Wednesday." U.S. businesspeople should view the CSFR as a long-term market, requiring a substantial time investment before a profit may be realized.

Embassy comments do not contemplate a personality change for each of the several hundred varied U.S. business visitors we counsel each month. We do plan, however, to relay these general observations to them, while reaffirming our general support and offer of direct assistance.

IV. FINANCIAL ENVIRONMENT

A. Currency and Foreign Exchange

On January 1, 1991, the CSFR government instituted internal convertibility of the crown for investment and commercial transactions, contained in the Foreign Exchange Act of November 28, 1990. This means that now importers and investors can freely buy foreign currency for current account transactions. Investors who are CSFR legal persons may convert profits earned in crowns to other currencies. Foreign investor companies are required, however, to exchange hard currency earned in trade to state banks for crowns. The banks must make hard currency available to these companies for profit repatriation or to pay for imports. These conditions do not apply to the initial investment in a joint venture, which may be held in a hard currency account, and other exceptions may be granted.

State bank (CSFR Central Bank) officials inform us that under the new foreign exchange law, payments for imports requiring less than the equivalent of six million crowns (200,000 dollars) in hard currency, will be made immediately. Payments requiring more than six million crowns in hard currency will normally take three months. Importers must borrow from the bank (one year credit) for purchases of six to 15 million crowns (up to 500,000 dollars); a two year credit of 15 to 30 million crowns (up to 1,000,000 dollars); and a four year credit for imports over 1,000,000 dollars. Current interest rates are 15 to 18 percent.

Repatriation of crown profits in hard currency should be possible in no more than a few days, although no precedent exists as yet for this type of transaction. Currency transfer rights apply to foreign companies which are CSFR legal entities, and to foreign representative offices. The state bank's role in the transfer process is limited to guarantor for the availability of hard currency. Actual transfers are made through one of the country's commercial banks, and no special permission is required from the state bank.

Foreign banks are being licensed to operated in the CSFR. U.S. Citibank, Societe Genrale (French), Creditanstaldt (Austria), and Landerbank (Austria) offer banking services in Prague at this time. There are 33 additional representative offices of foreign banks in the CSFR. Some of these banks are actively promoting investment financing in the CSFR.

B. Foreign Debt

Federal finance minister Vaclav Klaus said the CSFR's foreign debt would rise in 1991 from 7.6 billion dollars to around 9 billion dollars. Foreign loans, primarily from the IMF and EEC, will be used to stabilize the currency.

C. Future of Countertrade

The Federal Ministry of Foreign Trade reports that countertrade with Western countries will be discouraged, and will only be actively continued with traditional trading partners in the former Eastern Bloc. The Czech Republic Ministry of Economic Policy, however, promotes countertrade as a mechanism in large-scale enterprise privatization.

D. Export Controls

The CSFR government passed a law December 5, 1990 to safeguard the export of technology from the CSFR to third countries. The law went into effect February 1, 1991. This law paves the way to more liberalized export control of high-tech equipment to the CSFR. Now, in some cases high-tech equipment—except for the most sophisticated computers/military hardware—may be exported to the CSFR.

E. Bilateral and Multilateral Sources of Financing for Investment and Exports

1. U.S. Government foreign assistance funding for the CSFR in fiscal years 1991-92:
   - Privatization and Economic Restructuring: 45 million dollars for the CSFR/American Enterprise Fund (see below).
   - Banking and Finance: 2 million dollars for competition policy, laws and regulations; and bank training and technical assistance.
   - Environment: 15 million dollar sector grant for essential imports; 7 million dollars for technical assistance.
   - U.S. private voluntary organization initiatives: 3.5 million dollars for YMCA, American Jewish Distribution Center, Project Hope.
   - Energy Efficiency: 2.5 million dollars in technical assistance for industrial energy efficiency, refinery operation, improved international oil purchasing, and energy/price reform.

   Contact: USAID Washington, DC or Ms. Lee Rousell, USAID U.S. Embassy Prague (same address as US&FCS).

2. U.S./CSFR Enterprise Fund: A 60 million dollar bilateral assistance fund to promote the development of the CSFR private sector and to support the implementation of policies and practices conducive to such development. This fund is designed to provide loans, grants, equity investments, feasibility studies, technical assistance, insurance guarantees,and other help to a variety of small and medium-sized businesses, agricultural concerns, and U.S./CSFR joint ventures. The fund was legally established during Spring, 1991 with directors in the U.S. and the CSFR. Contacts: Prague: Mr. Peter J. Tischler, Blamicka 28/1008 120 00 Prague 2, Telephone: 011-42-2-25-41-30; Fax: 011-42-2-20-71-18. Bratislava: Mr. Leighton Q.J. Klevana, Radlinskeho 27 Quattro P.O. Box 66/1 810 05 Bratislava 15, Telephone: 011-7-565-44; Fax: 011-7-575-80. In Washington, DC, contact the fund at 1620 Eye Street, NW, Room 703, Washington, DC 20006.

3. The U.S. Trade and Development Program (TDP) is an independent U.S. government agency which promotes U.S. exports for major development projects in middle-income and developing countries. TDP funds feasibility studies, consultancies, training programs, and other project planning services related to major products. Contracts funded by TDP grants must be awarded to U.S. companies. U.S. involvement in project planning helps position potential U.S. suppliers of goods and services for follow-on contracts when these projects are implemented. TDP has been very active in the CSFR to date in the Environmental, Telecommunications, Aerospace, Transportation, Information Management, and Food Processing/Packaging Industries. Contact: U.S. Trade and Development Program, SA-16, Room 309, Washington, DC, 20523-1602; Telephone: 703-875-4357; Fax: 703-875-4009.

4. The U.S. Export-Import Bank promotes the export of U.S. goods and services through a variety of loan, guarantee, and insurance programs. All of its programs are available for the CSFR. Contact: EximBank, 811 Vermont Ave., NW, Washington, DC 20571; Telephone: 202-566-8990 or 800-424-5201 (Small Business Advisory Service).

5. The Overseas Private Investment Corporation (OPIC) is a self-sustaining, U.S. government agency which promotes growth in developing countries by encouraging U.S. private investment. OPIC's key programs are its loan guaranties, direct loans, and political risk insurance. For Eastern European countries, OPIC also offers an Eastern European Growth Fund, designed to match OPIC funds with private venture capital to finance new business; the small business loan guaranty program; and an environmental investment fund. Contact: OPIC, 1615 M Street, NW, Washington, DC 20527, Ms. Sue Goeransson; Telephone: 202-457-7183.

6. The World Bank (IBRD) and International Finance Corporation (IFC) are planning technical assistance loans for Eastern Europe. This money will focus on environmental infrastructure, telecommunications, and transportation infrastructure. Among the technical assistance loans would be monies for an enterprise fund to help local CSFR firms negotiate joint ventures or restructure. Contact: Mr. Francis K. Vita, Senior Country Operations Officer, The World Bank, 1818 H Street, NW, Washington, DC 20433; Telephone: 202-473-2970; Fax: 202-477-0710; Telex RCA 248423; and Mr. Vikas Thapar, Chief of Mission, International Finance Corp., Siroka 5 Prague 1, CSFR; Telephone: 42-2-232-5430; 232-5472; 232-5494; Fax: 42-2-232-5533.

7. The Multilateral Investment Guarantee Agency (MIGA) is part of the World Bank Group. Its purpose is to encourage foreign investment in developing countries by providing investment guarantees against the risks of currency transfer, expropriation, war and civil disturbance, and breach of contract by the host government; and advisory services to developing member countries on means of improving their attractiveness to foreign investment. Contact: Mr. Leigh Hollywood, Vice President, Guarantees, Multilateral Investment Guarantee Agency, 1818 H Street, NW, Washington, DC 20433; Telephone: 202-473-6168; Fax: 202-477-9886; Telex RCA 248423.

8. The European Bank for Reconstruction and Development (EBRD), a new multilateral institution, is expected to commence operations in London during Spring, 1991. The bank's capitalization will be approximately 12 billion dollars, to support annual lending of about 2.4 billion dollars during the first five years of operation. EBRD will support private sector development, Environmental Clean-up, and Infrastructure Development in Eastern Europe. Because of the U.S. 70 million dollar subscription to the bank, U.S. companies are guaranteed the right to bid on all projects and studies financed by it. Contact: S. Stanley Katz, EBRD, 6 Broadgate, London EC2M 2QS, U.K.; Telephone: 071-496-0060; Fax: 071-638-2375, 628-6088; Telex: 881-2161.

9. European Economic Community (EEC) funding for environmental projects in Eastern Europe: the EEC allocated about 39 million dollars for the CSFR in 1990. Funding levels for 1991 have not been decided. Contact: mr. Philippe Bourel, Division for International Affairs, Directorate General for Environment, Nuclear Safety, and Civil Protection (DG-XI), Commission of the European Communities, Room 10/248, 45 Avenue D' Auderghem, 1040 Brussels; Telephone: 32-2-235-3851; Fax: 32-2-235-5665.

V. TRADE AND INVESTMENT ISSUES/BARRIERS

1. Major Trade Barriers

A. Access to hard currency for import payments: although the CSFR now has full internal convertibility and may move to unrestricted convertibility within several years, access to hard currency for import payments remains problematic. Under the state bank, payments for imports valued at over 200,000 dollars will normally take three months. Importers must make a one year loan for amounts between 200,000 and 500,000 dollars; a two year loan for amounts between 500,000 and 1,000,000 dollars; and a four year loan for imports valued at over 1,000,000 dollars. While banking laws will be liberalized in early 1992, it remains unclear how this practice may be affected.

Host Government Position: conserving hard currency. Regulations of foreign currency loans.

USG Action: Working through multilateral institutions like IMF and IBRD to support the CSFR's economic stabilization. The CSFR has a manageable level of foreign debt.

B. Increase of Overall Tariff Base from 5 to 5.7 Percent

Host Government Position: Protect sensitive industries as the economy moves toward a market-oriented base. Also, to move away from the Communist Era System of controlling imports largely by limiting access to foreign exchange.

USG Action: We are opposed to this measure in the Gatt. We will work with the CSFR in the Gatt to lower tariff rates.

C. Increase of 15 Percent Import Surcharge on Consumer Goods and Foodstuffs for Sale in the CSFR to 20 Percent, Effective Indefinitely: Surcharge of 15 percent on imports of private articles for personal use (duty-free limit of 100 dollars).

Host Government Position: Conserve foreign exchange reserves.

USG Action: Support the CSFR in multilateral organizations, help it to attract investment and build its export-oriented industries.

2. Major Investment Barriers

A. CSFR government inexperience in procurement and international tenders results in procrastination and undue delays in contractor selection. This costs U.S. firms millions of dollars in travel and lodging expenses. Regulations should be implemented by the host government establishing specific deadlines for tenders and final procurement; and prohibition of new offer acceptances after the established deadline.

Host Government Position: Inexperience and hesitancy in making major decisions.

USG Action: Working with the national and local governments to help them improve procurement practices.

B. Inadequate protection of U.S. intellectual property: the CSFR is a party to the Bern, Paris, and Universal Copyright Conventions. While the CSFR record on protection of intellectual property has generally been good, the U.S. government is aware of one trademark violation an done patent violations which are currently under negotiation between U.S. firms and alleged CSFR violators. Both cases involve alleged violations over long periods of time. FCS notes some concern by potential investors, notably in the pharmaceutical and recording industries, that current intellectual property laws are inadequate protection for investment in local manufacturing.

Host Government Position: Passing new laws, strengthening regulations.

USG Action: Work with federal government to strengthen/enforce the trademark/patent/ copyright laws.

VI. MARKET ANALYSIS PLAN (MAP)

1. Industry Subsector Analyses (ISAS)

During FY 1992 US&FCS Prague is expected to complete at least 10 ISAS. Below are 10 subsectors which we plan to study. Researchers will be in-house, but all cannot be identified until staffing is completed (January, 1992 target date).

TOP 10 INDUSTRY SUBSECTORS:

1. A. Toxic Waste Treatment Materials/Equipment
   B. (POL)
   C. Due Date: Jan. 31
   D. CSFR
   E. Researcher: Emiliana Ulrychova
   F. Workdays for Research: 5
   G. Justifications: Serious toxic waste disposal problems; already many existing sites especially in Northern Bohemia Brown Coal Basin and at former Soviet military bases. Sources of financing: AID, IBRD, OPIC, EBRD.

2. A. Air Pollution Control Equipment
   B. (POL)
   C. Due date: Feb. 28
   D. CSFR
   E. Researcher: Emiliana Ulrychova
   F. Workdays for Research: 5
   G. Justification: Serious air pollution nationwide caused by burning soft coal for energy generation. Acid rain casing widespread environmental damage.

3. A. Fiber Optic Transmission Equipment
   B. (TEL)
   C. Due Date: June 30
   D. CSFR
   E. Researcher: ?
   F. Workdays for Research: 10
   G. Justification: After implementation of cellular system and packet switching network, CSFR will be examining fiber optic network to link all of Eastern Europe.

4. A. Electric Generating Equipment
   B. (ELP)
   C. Due Date: Mar. 31
   D. CSFR
   E. Researcher: ?
   F. Workdays for Research: 10
   G. Justification: Electric generating capacity needs to be expanded to meet increasing industrial and consumer demand; and the switch away from coal to other sources of energy.

5. A. Fast Food Franchises
   B. (FRA)
   C. Due Date: Feb. 28
   D. CSFR
   E. Researcher: Jiri Frantal
   F. Workdays for Research: 5
   G. Justification: Great opportunities exist to satisfy burgeoning tourist and local demand. Virtually none currently exist in market.

6. A. Financial Services
   B. (FNS)
   C. Due Date: April 30

D. CSFR
E. Researcher: Jiri Frantal
F. Workdays for Research: 5
G. Justification: Banking sector in transition, and emergence of private and foreign banking operations. All financial services are antiquated and extremely slow.

7. A. Travel and Tourism Services
B. (TRA)
C. Due Date: April 30
D. CSFR
E. Researcher: Emiliana Ulrychova
F. Workdays for Research: 5
G. Justification: 50 million plus tourists expected per year. Current infrastructure cannot begin to adequately support demand.

8. A. Management Consulting Services
B. (MCS)
C. Due Date: May 31
D. CSFR
E. Researcher: ?
F. Workdays for Research: 5
G. Justification: Privatization of enterprises creates need for company financial/accounting management, plant management, marketing, etc.

9. A. Hotel Automation
B. (HTL)
C. Due Date: June 30
D. CSFR
E. Researcher: Emiliana Ulrychova
F. Workdays for Research: 5
G. Justification: Existing and new hotels need help to better serve tourists/business visitors and to cope with increasing traffic.

10. A. Commercial Aviation Maintenance
B. (AVG)
C. Due Date: June 30
D. CSFR
E. Researcher: Jiri Frantal
F. Workdays for Research: 5
G. Justification: Privatization of airlines and leasing of new fleet of passenger planes; local production of new commuter aircraft; dramatic growth of air traffic.

VII. TRADE EVENT PLAN

A. Event: Scientific/Lab Instruments—Process Controls IOGA Trade Mission (Lab)
Dates: January, 1992
Location: Prague and Bratislava
Participation: Infrequent Exporters/NTM

B. Event: Computer Software IOGA Trade Mission (CSF)
Dates: March, 1992
Location: Prague and Bratislava
Participation: Infrequent Exporters/NTM

C. Event: National Machine Tool Builders' Association (NMTBA) IOGA Trade Mission
Dates: March 5-8, 1992
Location: Prague
MTL
Contact: Richard W. Kuba, International Marketing Director NMTBA
7901 Westpark Dr., P.O. Box 1312 McLean, VA 22102-4269
Telephone: 703-893-2900; Fax: 703-893-1151
Participation: NTM

D. Event: Saint Lawrence Seaway Development Corp. IOGA Trade Mission
Dates: March 14-17, 1992
Location: Prague
Maritime Services
Contact: Stephen J. Rybicki, Director of Marketing, U.S. Dept. of Transportation,
Saint Lawrence Seaway Development Corp.
Suite 5424, 400 Seventh Street, SW, Washington, DC 20590
Telephone: 202-366-0091; Fax: 202-366-7147
Participation: NTM

E. Event: Manufacturing Jewelers and Silversmiths of America, Inc. (MJSA) IOGA Trade Mission
Dates: to be determined
Location: Prague
JLR
Contact: Mr. David Rocha, MJSA
100 India Street, Providence, RI 02903
Telephone: 401-274-3840; Fax: 401-274-0265
Participation: Infrequent Exporters/NTM

F. Event: USA Tech '92 Capital Goods Solo Fair
Dates: April 27-May 1, 1992
Theme: Regional Capital Goods (U.S. only)
Location: Bratislava
Wash. Recruiter: Joseph W. Harrison
Telephone: 202-377-5455; Fax: 202-377-8219
Participation: Infrequent Exporters

G. Event: Insurance Business Development IOGA Trade Mission
Dates: May, 1992
Location: Prague and Bratislava
INS
Participation: NTM

H. Event: High-Tech Europe Catalog Show
Dates: June, 1992
Location: Prague
CPT, CSF, TEL, TES
Contact: Matthew Edwards, USDOC/ITA/USFCS/EPS
Telephone: 202-377-0584; Fax: 202-377-0178
Participation: Infrequent Exporters/NTM

I. Event: Danube Basin Conference
Dates: June 1-2, 1992
Location: Vienna, Prague
Horizontal small/medium sized business
Contact: Benjamin Brown, SCO, USFCS AMEmbassy Vienna, Austria
Participation: Infrequent Exporters/NTM/NTE

J. Event: International Engineering Fair
Dates: September, 1992
Theme: Industrial, Technical Equipment
Location: BRNO
Participation: New, Infrequent and Experienced Exporters

K. Event: U.S./CSFR Joint Commercial Commission
U.S. companies with an interest in doing business in the CSFR should contact the U.S. Department of Commerce in Washington, DC; East European Business Information Center; Telephone: 202-377-2645, Fax: 202-377-4473; CSFR Desk Officer Michael Mowrey; Telephone: 202-377-4915, Fax: 202-377-2155; 377-4098; or the U.S. & Foreign Commercial Service, U.S. Embassy Prague, Robert M. Shipley, Commercial Attache, Shelley Galbraith, Commercial Officer, 15 Trziste 125 48 Prague 1, or AmEmbassy, Unit 25402 APO AE 09213-5630, Telephone: 42-2-536641, 42-2-532-470, Fax: 42-2-532457.

VIII. WORK PLAN

Following is FCS Prague's work plan for FY 1992:

Critical Element I: Market Analysis and Program Planning (weight 20%)

CORE PROGRAM

1. Develop the country marketing plan and revise as necessary during the year.
2. Report host country trade barriers and trade and investment policies.
3. Assess the commercial and economic environment and provide host country trade data.
4. Provide alert reporting on significant commercial developments and quality market research, including ISAS.

Critical Element II: U.S. Government Commercial Program Support (weight 20%)

CORE PROGRAM

1. Provide program and visitor support to U.S. government agencies, including IEP, TD, IA, TDP, EximBank, OPIC, NOAA, PTO, USTTA, USTR, NIST, NTIA and BXA.
2. Carry-out the overseas export control responsibilities of the department, as identified in the ITA-BXA memorandum of understanding.
3. Support and participate in bilateral and multilateral trade negotiations.
4. Report on and support major project opportunities.
5. Provide support to host country government official visits to the U.S.

Critical Element III: Business Counseling and Outreach (weight 20%)

CORE PROGRAM

1. Maintain key host country government and business contacts.
2. Counsel and assist U.S. business visitors.
3. Advocate on behalf of U.S. business interests.
4. Support business activities related to licensing, joint-ventures, and export related investment.
5. Assist in resolving commercial trade complaints.
6. Serve as liasion with local business groups.

Country Specific Activities:
1. Organize and Inaugurate an AMCHAM.

Critical Element IV: Trade Promotion and Commercial Information Support (weight 30%)

CORE PROGRAM

1. Implement ADS, WTDR, TOP and CSS services.
2. Develop and carry-out a market research program (ISAS, best prospects, market insight information, and alert reports).
3. Disseminate information in "Commercial News USA."

Country Specific Activities:
1. Develop and carry-out market-relevant trade promotion events.
2. Hire EEBIC contractor and begin sending trade/investment opportunities for "Eastern Europe looks for Partners."

Critical Element V: Management, Supervision and Administration (weight: 10%)

CORE PROGRAM

1. Manage program and personnel resources.
2. Manage administrative and budget requirements.
3. Report on event receipts, expenditures, end-of-show results, and event scheduling changes.
4. Prepare performance plans and conducts midyear progress reviews and annual performance appraisals.
5. Initiate the commercial information management system (CIMS) and train staff on PC operation and programs with the assistance of FCS Washington and Rac Vienna.
6. Plan for and provide training opportunities.
7. Participate in and support country team activities.
8. Ensure that the department's policies and requirements regarding the EEO program are carried-out at post.

Country Specific Activities:

1. Bring staff to full capacity—2 Americans, 6 FSNS, 2 full-time PSCS, one PSC driver and 1 USDA/FAS FSN.
2. Move into new FCS commercial center offices.
3. Secure more ADP equipment from FCS Washington.
4. Order core publications for a new business reference center.

APPENDIX M - A NEW ERA BEGINS FOR CENTRAL AND EASTERN EUROPE

This appendix was made available by the U.S. Department of Commerce, 14th Street and Constitution Avenue, NW, Washington, DC 20230. It was originally published as an article in Business America, volume 113, number 10, May 18, 1992.

Remarks by Secretary of Commerce Barbara Hackman Franklin at the Opening Plenary Session of the West-East Economic Summit Muenster, Germany May 8, 1992

Good afternoon. I am so pleased to be here today. As you know, this is my first opportunity to join in a meeting with all of you—my colleagues from the other leading democracies of the world and from the newest democracies as well. And I am delighted to meet all of you.

I am also pleased about the topic we are here to discuss. This is such a dramatic and historic time in the world. Together, our countries have pursued a course of action that has brought peace and the hope for prosperity to all the world. Democracy has prevailed. New free market economies are springing up everywhere.

Now the challenge is to stabilize these gains. Great progress has been made. Good will and good intentions prevail. We must be sure that we build upon this promising start; that we continue on a path that expands world peace and prosperity.

Cooperation and determination have brought us to this stage today. Continued cooperation and determination will help us achieve our vision for tomorrow.

Certainly the new democracies of Central and Eastern Europe and of the former Soviet Union are doing their part. The courage and determination of the people of these countries has inspired us all. They have withstood attempts to turn back the clock and stop the progress of democracy. They are bravely confronting hard economic realities, stoically surviving hard times and putting their faith in a better tomorrow. They are transforming their economies, and undertaking meaningful reform.

We must help them realized this vision of tomorrow. Building strong, robust free market economies in Central and Eastern Europe, in Russia, the Ukraine and Kazakhstan as well as the other new states adds to global security and offers the tantalizing possibility of increasing the standard of living for all people on earth.

We in the United States will do our part. President Bush has pledged this. He has laid out a course of action to do so. We are working actively to remove trade and investment restrictions left over from the Cold War. We are providing technical and financial assistance. And the United States Congress is now considering a far-reaching proposal, called the Freedom Support Act, which the President has submitted to our Congress to further expand these activities.

But perhaps the most important part of President Bush's plan to support the development of free market economies in this vital part of the world is to smooth the way for the growth and expansion of the private sector. This, of course, is why we are here—to consider ways in which our countries can support and encourage private foreign investment in these emerging democracies.

I am heartened by what I heard this morning from the business representatives who accompanied us. Clearly, our private sectors are very interested in economic possibilities offered by these new countries and are very willing to explore investment opportunities.

I know this is true in the United States. Our friends who spoke this morning said it. And I know it from my own experience. As you may know, I just recently left the private sector to serve in my government. For the past decade, I was a member of the boards of directors of seven large American corporations. And I heard these issues discussed. In fact, I was an advocate for supporting these developing free market economies. I know firsthand how interested American businesses are in investing in this region of the world.

But we are also hearing it from all parts of the American business community. Large corporation, small- and medium-sized businesses, manufacturers and service companies—all interested in exploring the opportunities in these new free market economies.

But they have concerns. Many of these concerns are identified in a report our Department of Commerce has just released. This report discusses the obstacles to trade and investment in the new republics of the former Soviet Union. It is based on comments and observations by more than 100 American companies which are trying to do business in these countries. We have a copy for each of you.

What this report shows is that overcoming decades of central-planning to create a robust private sector is no simple task. It requires time, effort, and cooperation among all nations.

In the view of the United States, one of the vital keys to success is the privatization process. So far, small-scale privatization seems to be proceeding well in a number of countries. Unfortunately, larger-scale efforts are not achieving as much success.

Experience in Central and Eastern Europe underscores the truth of this observation. Privatization has been most successful when applied to medium- and small-sized business activity. Encouragement and stimulation of this type of activity in Eastern Europe and the states of the former Soviet Union should be a goal of this conference. The vitality and creativity of small- and medium-sized businesses will be as essential to the growth of the East European and NIS economies as it is to growth and creation of jobs in market economies everywhere. The same entrepreneurial spirit that has powered economic growth in our own countries and other parts of the world can do so as well in this region.

Another key to private sector development and overall economic growth is foreign private investment. While foreign investment will not solve all the economic problems confronting these states, it is vital to long-term economic growth and development throughout the region.

Foreign investment contributed much to creation of free market economies. It creates capital inflow. It helps introduce new and necessary technologies that spur economic growth. It provides needed jobs. And it helps develop an export base vital to commercial success.

But the global competition for investment capital is intense. The new democracies of the region will have to compete aggressively for investment funds, not just among themselves, but also with the countries of Asia, Latin America, the Pacific Rim and Africa. Many of these nations have taken aggressive action as well to attract foreign investment.

This means that the new democracies will have to remove obstacles to foreign investment. In particular, they should act quickly to fully normalize their bilateral trade relationships with established market economies and with each other.

I also urge my colleagues from these states to remove or avoid erecting tariff or other barriers to trade among themselves. They also should investigate the possibilities of creating regional, GATT-consistent free trade areas. These areas can be a stimulus to the efficiency and innovation necessary for growth of internationally competitive industries.

The United States applauds the efforts of Poland, Czechoslovakia, and Hungary toward establishing a free trade area. This will have a significant effect on attracting investment to the region.

Another step these nations must take to attract foreign investment is to provide reliable information. This must be a high priority. Business people base investment decisions on reliable information ranging from official economic statistics to the creditworthiness of individual enterprises. The absence of credible data is a major impediment to investment.

Another great source of frustration for U.S. investors remains the uncertainty surrounding aspects of the business climate in the reforming states. This includes conflicting laws and regulations, ill-defined property rights, and overlapping governmental jurisdictions.

As a long-term step toward building robust economies, our friends in the region must begin as quickly as possible to resolve the lace of adequate financial and legal systems. And they must implement fully convertible currencies. They also must address the problems of inadequate infrastructure, including transportation and communications systems.

We in the United States stand ready to help, just as do all of our friends who make up the G-7. These are exciting and challenging times. I am confident a wealth of ideas will emerge from this conference as to how we can assist the brave people of these countries achieve their dreams of a brighter tomorrow.

I propose that this Muenster Conference issue a Declaration of Principles on Investment Activity. This declaration can be used as a guidepost by the newly democratic and free-market as by our deliberations here in Muenster.

Such a declaration gives concrete meaning to our efforts here in Muenster. It also will encourage Western companies to explore the promising new investment markets of the region and helps the countries of Eastern Europe and the NIS as they transform themselves into viable commercial partners in the world trading system.

We have and unprecedented opportunity today to create a new world in which dreams of peace and prosperity become the reality for all people. Working together, committing ourselves and our countries to striving for this reality, we can reach for the stars—we can achieve the stars.

Summary of Conclusions of the West-East Economic Conference Of Ministers of Economics, Industry and Trade In Muenster, Germany, May 7-9, 1992

At the invitation of the Federal Minister of Economics of the Federal Republic of German, the First Deputy Prime Minister of Balarus, the Prime Minister of Kazakhstan, the Vice Prime Minister and Minister of Economics of the Ukraine, and the Ministers of Economics, Industry and Trade of Canada, the Czech and Slovak Federal Republic, France, Hungary, Italy, Japan, Poland, Portugal, the Russian Federation, The United Kingdom, and the United States of America, as well as representatives of the Commission of the European Communities, the Organization for Economic Cooperation and Development (OECD), and the European Bank for Reconstruction and Development (EBRD) met in Muenster from 7 to 9 May 1992.

Giving special attention to the pivotal role of the private sector in economic reform, they exchanged views on policy measures and possible assistance to help accelerate the transition of the countries of Central and Eastern Europe as well as those of the former Soviet Union to market economies, and, in particular, necessary conditions for closer international business cooperation. Representatives of industry from participating countries took part and made valuable contributions based on their experience to date in the reform economies.

The participants agreed on the vital importance of encouraging rapid development of the private sector in the reforming countries and of accelerating foreign direct investment, which can play a major role in the economic reform process. They examined necessary policy measures by the reform states and addressed various areas of useful assistance by established market economies, including privatization, the creation of a small- and medium-sized enterprise sector, conversion from military to civil production, enhancement f the role of trade, training in market economy skills, cooperation in environmental protection, and in particular, framework conditions that are necessary for investment from the private sector. Participants also acknowledges the vital role of the international financial institutions and other specialized organizations in this context.

The participants expressed strong support for the reform efforts being undertaken. They recognized that the process of transition to market economies has advanced to different degrees among these countries and that some, in particular the Czech and Slovak Republic, Hungary, and Poland, may well be in a position to share experience with and offer advice and technical assistance to others that are beginning the reform process, above all as regards privatization. Participants also acknowledged that experience taken from established market economies may have to be adjusted when transferred to reform countries and that realistic time horizons should be assumed to take account of the differences featured by the reform countries both as a group and individually. In a bilateral learning process between established market economies and reforming states, creative solutions are called for.

Participants emphasized that the performance of market-based economies relies primarily on the freedom of individual enterprises to make their own decisions, while governments provide the overall framework for economic activity. The provision of a stable macro-economic framework ranks high among the necessary conditions for private economic activity.

REFORM OF ECONOMIC STRUCTURES

There is an urgent need to reform economic structures, e.g., to implement demonopolization and privatization which will promote management automony, entrepreneurial initiative, and accountability. Even while enterprises remain state-owned, breaking up monopoly enterprises will actively promote competition. The established market economy countries and some of the more advanced reform countries agreed to provide, and, if possible, extend help and advice on privatization and on any mechanisms they had adopted to regulate monopolies. Participants also agreed on the importance of fostering the establishment of a healthy small- and medium-sized business sector. The established market economies offered to share their experiences in supporting small- and medium-sized businesses, including specialized government agencies, an appropriate legal framework, supporting financial institutions, training centers, and advice on business start-ups. Participating countries agreed to encourage and facilitate the establishment of chambers of industry and commerce, small business consortia, trade associations, and similar bodies in the reform countries and to foster links with their equivalents in the established market economies.

CONVERSION OF MILITARY PRODUCTION

The conversion of military production to civilian uses is of key importance to help market-based structures evolve. Success in the conversion process requires the definition of a clear economic strategy integrating structural, institutional and microeconomic approaches. The high concentration of such military facilities in some areas has important implications for regional economic development. International cooperation at the corporate level could expedite conversion. The full range of information should be provided to enable prospective partners to assess the economic potential of military enterprises. Financial and ecological burdens inherited by these enterprises also have to be addressed.

EXPANDING INTERNATIONAL TRADE

All countries agreed on the importance of trade and place a high priority on the successful conclusion of the Uruguay Round of Multilateral Trade Negotiations. AS part of this process the reform states accepted the need to assume the obligations and disciplines of the international trading system.

Participants stressed the importance of market access, including in sensitive sectors, for the exports of the reform countries and the value of measures such as trade preferences and regional free-trade agreements. The representatives of the reforming economies agreed on the need to continue traditional trade relations on a free-market basis, to avoid the introduction of new trade barriers against each other, and to look for ways to raise the level of trade within the region. Those who are not already Contracting Parties will seek to participate in the GATT system as quickly as possible.

To advance trade in general, problems of real pricing, functioning foreign exchange mechanisms, the lack of efficient banking operations, and delays in external payments all have a bearing and need to be solved by the reform states.

In implementing current relief programs, distortion of competition should be avoided and due consideration should be taken of commercial interests of trading partners in the region. The established market economies offer specific advice and assistance in building institutions that support trade, e.g., setting up export credit insurance and general trade promotion schemes, in marketing and in training customs officials.

PROMOTING PRIVATE INVESTMENT

The participants agreed on the vital importance of contributions that private investments from established market economies will have to make to the reform process. To this end, framework conditions for private investors and the general conduct of business must be further improved and adapted to international standards and practice.

Drawing from the comments made by participants from industry, some key areas were identified for stimulating private investment, and participants agreed that the application of the following set of principles and recommendations would improve investor confidence and interest in the reform states:

- ensuring a positive investment climate in which investment flows are determined by market forces with fair and equitable non-discriminatory treatment of foreign investors not less favorable than domestic competitors, avoidance of performance requirements, assured free transfers related to investments and fully convertible currencies;

- developing a secure and stable legal framework through adequate protection of investments including effective intellectual property rights, access to third-party arbitration, the effective enforcement of contracts, and guarantees of prompt and adequate compensation in case of expropriation.

- facilitating market access through the dissemination of reliable and comprehensive economic data, transparent and stable commercial laws and regulations; active promotion of business contacts and streamling admission procedures.

Fostering international cooperation, established market economies will seek to respond positively to requests for technical assistance and advice on matters such as legislation, procedures and the establishment and operation of bodies to promote inward investment. Responding to the reform efforts, insofar as applicable, they will also review their own incentives for investment. Responding to the reform efforts, insofar as applicable, they will also review their own incentives for investment, such as existing capital investment guarantee systems and other measures, including those for small- and medium-sized enterprises.

TRAINING

Participants from the established market economies agreed to encourage and, where possible, assist their enterprises to offer training positions for managers, supervisors and other staff in enterprises from the reform states. The governments of the reform states could establish centers for disseminating modern management and technological know-how with advice and technical assistance by the established market economies.

ENVIRONMENTAL PROTECTION

The transition process offers the challenge of simultaneously reforming the industrial fabric and overcoming environmental problems. Participants agreed that governments in the reform states, with advice and technical assistance from established market economies, should aim to set adequate environmental standards in line with international levels, ensure standards in line with international levels, ensure enforcement and promote the application of environmentally advanced technologies. In this respect, participants acknowledged the work under way in appropriate international bodies. They emphasized the major role played by proper economic mechanisms and incentives and stressed the role of private sector investments which advance the introduction of cleaner technologies, together with the associated skills and know-how. They specifically pointed out that there is a need to clean up past environmental pollution by reform states in order to avoid creating additional investment barriers.

FOLLOW-UP

Participants agreed to inform non-participating reform countries of the discussions held at Muenster. They believed their meeting fruitfully addressed an area of special importance that deserves more attention in international efforts to provide assistance to reform countries. They agreed, therefore, to continue this consultation process and, together with business representatives, to review progress in the areas of cooperation identified at the Muenster conference within one year.

APPENDIX N - EASTERN EUROPE LOOKS FOR PARTNERS

This appendix was made available by the Eastern Europe Business Information Center, U.S. Department of Commerce, International Trade Administration

Under the Eastern Europe Business Development Program of the American Business Initiative, the Eastern Europe Business Information Center (EEBIC) has created the "Eastern Europe Looks for Partners" information service for readers of its *Eastern Europe Business Bulletin.* This service is designed to help U.S. businesses locate partnership opportunities in the rapidly expanding East European and Baltic markets.

The opportunities are provided solely as an informational service and do not represent an endorsement by the Department of Commerce. Verification of these leads is the responsibility of the reader.

If your company should decide to pursue one or more of these opportunities and would like our advice and assistance in doing so, please feel free to contact EEBIC at (202) 377-2645.

APPENDIX O - U.S. DEPARTMENT OF COMMERCE BUSINESS SERVICES IN EASTERN EUROPE

This appendix was made available by the Eastern Europe Business Information Center, the U.S. Department of Commerce, Room H-7412, Washington, DC 20230. It is dated 1992.

The U.S. Department of Commerce provides a variety of services to firms interested in trade and investment opportunities in Eastern Europe. Although not every service is currently available in every Eastern European country, in the near future most of the services will cover the region. A majority of these services are available through the U.S. Department of Commerce District Offices. A list of District Offices is attached.

Agent/Distributor Service (ADS). ADS service is available through your local District Office. The ADS service will help your firm enter the Eastern European market by verifying your product's marketability and searching for a qualified foreign representative. The service is now available in Czechoslovakia, Romania, Hungary and Poland for $125.

World Traders Data Reports (WTDR). WTDRs are available through your local District Office. WTDRs will help screen your potential Eastern European customers and prospective agents by providing thorough confidential reports on potential foreign trading partners including buyers, distributors, agents, and retailers. WTDR is now available in Poland, Hungary, Czechoslovakia and Romania for $100.

Comparison Shopping. Sandra Collazo, Room H-1517, U.S. Department of Commerce, Washington, DC 20230, 202-377-8972. This is a customized service that provides companies with individualized market information on sales potential, local suppliers, major sales channels, market prices and buyers needs. The program is now available in Hungary for $1,250, in Romania for $750, in Czechoslovakia for $1,250 and in Poland for $1,000.

Gold Key Service. Joan Edwards, Senior Commercial Officer, American Embassy Warsaw, APO NY 09213, Telephone: 4822-214515, Fax: 4822-216327. Gary Gallagher, Senior Commercial Officer, American Embassy Budapest, APO NY 09213-5270, Telephone: 36-1-1228600/1217, Fax: 36-1-1422529. Gold Key Service ensures that appropriate meetings with government members, firms, and individuals are set up prior to arrival. Gold Key is now available in **Poland** and **Hungary**. Costs for the service are $250 for 6 key contacts in your field and $150 per day for a full schedule of appointments.

Overseas Trade Fairs/Missions. For information on how to participate in 1992 trade fairs in Eastern Europe and a schedule of events, call EEBIC at 202-377-2645. Trade fairs are a cost efficient and easy way to enter a new and emerging market such as Eastern Europe. In some cases the Department of Commerce mounts its own exhibit, in others it works with private show organizers to arrange special services for U.S. exhibitors.

Overseas Catalog/Video Catalog Show. For information on catalog shows to Eastern Europe, contact EEBIC at 202-377-2645. Get exposure for your product or service without the high cost of travelling overseas. The Commerce Department sponsors catalog shows through its overseas offices. Using the firm's literature or promotional video, the material will be displayed to a select audience.

Export Counselling. Call your local District Office or EEBIC at 202-377-2645. Trade Specialists are available to answer questions on how to invest in Eastern Europe, sources for U.S. Government funding, and potential joint venture opportunities. EEBIC also serves as a referral point for programs of voluntary assistance to Eastern Europe and the Baltics.

APPENDIX P - CENTRAL & EASTERN EUROPE BUSINESS INFORMATION

This appendix was provided by the International Trade Administration, Eastern Europe Business Information Center (EEBIC). It is dated March/April 1992.

About the Eastern Europe Business Information Center (EEBIC)

RECENT BUSINESS INFORMATION AVAILABLE

EEBIC acquires publications directly from Central and Eastern European countries and makes them immediately available to the public through NTIS. EEBIC responds to the U.S. demand for information on business opportunities in Central and Eastern Europe. It assists small- and medium-sized companies interested in exporting or joint ventures and answers questions about financing, export licensing, and trade promotion opportunities. EEBIC also had information on foundations or individuals who provide assistance to Eastern Europe.

Inquiries should be directed to: EEBIC, Room 7412, U.S. Department of Commerce, 14th and Constitution Avenue, NW, Washington, DC 20230, (202) 377-2645, Fax: (202) 377-4473.

Central and Eastern Europe Legal Texts Service

SUBSCRIPTION SERVICE AVAILABLE

The foreign commercial laws and regulations listed in this catalog are also available on a subscription basis. As these laws and regulations are received by the Department of Commerce, you will automatically be sent copies based upon the country selection. While most texts are provided in English, texts received in the original language are made available promptly and in most cases are followed by the English translation.

Texts are available through a standing order subscription. Standing orders are issued as texts become available, but not at fixed intervals. As texts are mailed, the cost is charged against your NTIS deposit account (see below).

Legal text from	*Standing Order Number English Translation*	*Standing Order Number Original Language*
Albania	PB92-961400BCQ	PB92-961500BCQ
Bulgaria	PB92-960400BCQ	PB92-960500BCQ
Czechoslovakia	PB92-960200BCQ	PB92-960300BCQ
Hungary	PB92-960600BCQ	PB92-960700BCQ
Poland	PB92-961000BCQ	PB92-961100BCQ
Romania	PB92-960800BCQ	PB92-960900BCQ
Yugoslavia	PB92-961200BCQ	PB92-961300BCQ
The Baltic Republics		
Estonia	PB92-966200BCQ	PB92-966300BCQ
Latvia	PB92-961600BCQ	PB92-961700BCQ
Lithuania	PB92-961800BCQ	PB92-961900BCQ

An NTIS deposit account is required for any of these subscriptions. The price is based upon number of issues. The minimum deposit for one of these catagories is $150; the minimum deposit for all six countries in English or in the original languages is $500. To begin your standing order subscription, call the NTIS subscription desk at (703) 487-4630. A deposit account application is included on the order form in the back. A deposit is not required for placing orders for individuals texts.

APPENDIX Q - EASTERN EUROPE BUSINESS BULLETIN

This appendix was made available by the Eastern Europe Business Information Center, the U.S. Department of Commerce, International Trade Administration. It is dated July/August 1992.

Special Feature: ***Romania***

EASTERN EUROPE BUSINESS BULLETIN

July/August 1992

Published by the Eastern Europe Business Information Center • U.S. Department of Commerce • International Trade Administration

Romania Offers Improved Business Climate to U.S. Firms

In February of this year, democratic coalition candidates in local Romanian elections moved to a sweeping victory. As a result, international confidence in Romania's economic and commercial future has been growing steadily.

In the area of U.S.-Romanian trade, the marked improvements in commercial climate and bilateral relations are beginning to be felt. U.S. exports are expected to rebound this year from their significant decline in 1991, as Romania implements liberalizing commercial legislation and realizes a higher level of multilateral financing.

When the United States granted a Jackson-Vanik waiver to Romania in August 1991, it marked a positive step forward in bilateral trade relations. As a result, the U.S. Trade and Development Program now sponsors feasibility studies in the Romanian market and the U.S. Export-Import Bank provides short-term insurance coverage through the Foreign Credit Insurance Association. Overseas Private Investment Corporation insurance guarantees are not yet available.

A recently negotiated bilateral trade agreement which would grant Most-Favored-Nation (MFN) trading status to Romania has been signed and submitted to Congress for approval. In addition to granting MFN status, the agreement would facilitate U.S. commercial activity in Romania and provide national treatment for trade-related financial transactions.

Best prospects for U.S. exports to Romania include telecommunications equipment; agricultural equipment and chemicals; medical equipment; computers, peripherals, and software; food processing and packaging equipment; electrical power equipment; pulp and paper machinery; and automotive parts and service equipment.

In the area of U.S.-Romanian investment, the Romanian investment climate is steadily improving. A liberal foreign investment law allows 100 percent foreign ownership of an enterprise, full hard currency profit repatriation, and generous tax incentives. A National Bank decree issued in November 1991 further elaborates that foreign investors may convert and repatriate 100 percent of all after-tax profits and stipulates that investor accounts will be in Romanian lei. Although Romania's scarce foreign exchange reserves and primitive banking system will continue to constrain the process of capital repatriation, technical assistance from multilateral sources should lead to improvement.

On May 28, 1992, a bilateral investment treaty ensuring national treatment guarantees for American and Romanian investors was signed. Following Senate ratification, the treaty would provide intellectual property right protection and a dispute resolution framework.

Over 550 joint ventures with American companies are registered in Romania, totaling $43.7 million in invested capital. A majority of these investments are small operations in the trade services field. Sectors with excellent investment potential include telecommunications, tourism, agribusiness, infrastructure development, construction, and light industry.

The Bucharest International Trade Fair is an excellent way to establish business contacts and get an overview of the Romanian market. Each fall the Commerce Department sponsors a U.S. pavilion at the Fair. In addition, a Commerce trade mission featuring food processing equipment and technology is planned for Fall 1992.

For an update on the status of U.S.-Romanian bilateral relations, call Romania Desk Officer Lynn Fabrizio on 202-377-4915. For further information on business opportunities in Romania, contact Commerce's Eastern Europe Business Information Center at 202-377-2645.

The Annual Bucharest Fair Is Largest in Romania

The Bucharest Fair, Romania's major trade fair covering both consumer and industrial product categories, provides U.S. firms with an opportunity to locate new business prospects, renew old contacts, and enhance their exposure in the East European market. Over 500 Romanian firms participated in last year's fair.

The U.S. Department of Commerce sponsors a pavilion for U.S. firms wishing to participate in the Bucharest Trade Fair. The participation fee for a booth includes space rental, design and construction of exhibit areas, an intensive market promotion campaign, inclusion in the U.S. exhibition brochure, and the full support of Embassy Commercial staff.

Last year, a record number of U.S. companies participated in the U.S. pavilion. Forty U.S. firms reported over 1,000 sales leads, 42 agent/distributor agreements, three joint ventures, and well over $3.5 million in export sales. This year, trade with Romania looks even more promising with trade agreements being finalized and MFN status waiting to be ratified by the U.S. Congress.

Best prospects for U.S. exporters include telecommunications equipment; computers, peripherals, and software; food processing and packaging equipment; electrical power equipment; pulp and paper machinery; and automotive parts and service equipment.

For more information, contact:

Kay R. Kuhlman
Senior Commercial Officer
American Embassy Bucharest
APO NY 09213-5260
Tel: 40-0-10-40-40
Fax: 40-0-12-03-95

APPENDIX R - NATIONAL TRADE DATA BANK

This appendix was made available by the U.S. Department of Commerce, Economics and Statistics Administration, Office of Business Analysis.

The NTDB was established by the Omnibus Trade and Competitiveness Act of 1988 to provide "...reasonable public access, including electronic access..." to export promotion and international economic information. The NTDB is your one-stop source for the best international trade and export data from 15 U.S. Federal Government agencies.

The NTDB brings you this wealth of information using the latest in information technology—Compact Disc-Read Only Memory (CD-ROM). Each month's disc contains the complete Data Bank and two software programs (BROWSE and ROMWARE®) to simplify access and retrieval of text, time series and tabular data. BROWSE was created exclusively for the National Trade Data Bank and was designed to help users with virtually no knowledge of Federal data explore and access the entire NTDB. Both programs perform databank searches based on country and product information, or searches for specific words in the titles of documents. Retrieved information can be viewed on screen, printed or exported to an ASCII or spreadsheet file.

You can use the NTDB CD-ROM on any IBM-compatible PC with at least 640K of memory, an ISO 9660 (standard) CD-ROM reader, and Microsoft CD-ROM extensions (version 2.0 or higher). The monthly NTDB CD-ROM is entirely self-contained and requires no additional space on a hard drive. NTDB software will optionally use a color monitor (CGA, EGA or VGA) and a mouse.

At $360 for a 12-month subscription or $35 per single disc, the NTDB is an economical means of locating the research data and overseas contacts essential to your success in exporting.

To order, simply call the NTDB staff at 202-377-1986.

APPENDIX S - EEBICFLASH DOCUMENT LISTINGS

This appendix was made available by the Eastern European Business Information Center, U.S. Department of Commerce, Room H-7412, Washington, DC 20230. It is dated 1992.

This is the Department of Commerce's Eastern Europe Business Information Center EEBICFLASH 24 hour document hotline. Simply call (202) 377-5745 from a touch-tone phone (don't touch your fax machine) and follow the ordering instructions. All you will have to do is enter the document order numbers listed below and your fax number when asked, and the documents automatically will be faxed to you in about five minutes. You may order up to five documents per call. This listing is updated on the first of every month.

Order Number	Document
I. Export Information	
1010	Eastern Europe Business Bulletin Subscription Form
1020	Eastern Europe Business Information Center (EEBIC) Description
1025	U.S. Department of Commerce Services in Eastern Europe
1030	U.S. Department of Commerce Offices in Eastern Europe
1035	East European Commercial Offices in the U.S.
1040	Eastern Europe Laws (in English) Ordering Information (NTIS)
1045	Consortia of American Businesses in Eastern Europe (CABEE)
1050	U.S. Financial Institutions Active in East-West Business
1055	U.S. Trade with Eastern Europe
1060	U.S. East European Economic Councils
1065	American Business Center in Warsaw
1110	U.S. Department of Commerce (US&FCS) District Offices Business Counseling Services
1111	U.S. Department of Commerce District Offices Addresses and Phone Numbers
1115	Agent/Distributor Service
1120	Export Contact List Service
1125	U.S. Department of Commerce Gold Key Appointment Service
1130	Comparison Shopping Service To Find & Asses Your Markets
1135	Catalog & Video/Catalog Exhibitions Export Promotion Service
1140	Foreign Buyer Program (export trade fairs in the U.S.)
1145	Trade Shows To Promote Your Products
1150	Matchmaker Trade Delegation Service
1155	Trade Missions To Promote Your Products
1160	Foreign Market Research
1165	World Traders Data Reports
1170	The National Trade Data Bank Information Service
1175	Commercial News USA Export Promotion Service
1180	Trade Opportunities Program
1185	Electronic Bulletin Board Trade Opportunity Listing
1010	*Eastern Europe Business Bulletin* Subscription Form

Export and Investment Opportunities

2005	10 COMPANIES BEING PRIVATIZED BY AUCTION IN ROMANIA
2010	UN Peacekeeping Force in Croatiz Offers Major Sales Opportunity ($621m)
2011	$40 million for Agriculture, Power, Transport and Administration in Albania
2012	$32 million for Slovenia Water Project (water pipes and fittings)
2013	$227 million for Polish Medical Equipment, Software, Technical Assistance, and Training
2014	$187.2 million Civil Works, Goods, Consultant Services for Polish Enterprise Development
2015	$40 million for Agriculture, Energy, and Health Exports to Latvia
2016	Installation of Walls, Ceilings, Illuminators, and Air Conditioning Systems in Hungary
2017	$150 million for Equipment for Training and Research and Technical Assistance in Hungary
2022	Hungarian Shopping Center and Duty-Free Zone Development Project
2030	Needed Immediately: 2,000 Tons of Paper Per Month
2036	Tongs for Ingots, Round and Rectangle Bars for Slovenia
2041	Telecommunications Project ($84 million) in Czechoslovakia #1
2042	Telecommunications Project ($43 million) in Czechoslovakia #2
2045	Telecommunications Project in Romania
2047	Telecommunications Project in Latvia
2050	Heat Generation and Supply Facilities Project in Poland
2055	Harbor Construction and Waterborne Transport Projects in Czechoslovakia
2060	10-Story Office Building Construction in Budapest, Hungary
2065	Mine Utilization, Mine Reclamation, and Industrial Park Development in Hungary
2070	60% of Hungarian Wholesaler/Retailer Open for Bid
2071	Hungarian Real Estate Fund Manager Needed
2072	Privatization Consultant Needed for Budapest Bank
2073	$2.2 million of Digital Processing Unites and Related Software
2074	$557.5 million for Power and Environmental Project in Czechoslovakia
2075	Hungarian Customs Agency Seeks Joint Venture Partner
2076	$167.7 million for Agricultural Machinery, Inputs, Working Capital, Consultants in Romania
2077	Conveyor Belts and Electrical Cable for Lignite Mine Rehabilitation in Romania
2078	$17 million for Oil Field Equipment, Data Processing Equipment, Consultants in Romania
2079	Modernization of Heat and Power Station in Poland
2080	Nuclear Fuel Facility Environmental Impact Statement Needed in Czechoslovakia
2081	Fork-Lift Trucks (for 40-foot containers) for Hungary
2083	Estonia to buy $30 million of Agriculture, Transport, Energy, and Health Products by Oct.
2084	8,500 Crossply Tires needed in Romania: Bids due September 2, 1992
2085	Emergency Generating Sets and Power Supply System for Romania: Bids due September 17, 1992
2086	$115 million for Road Reconstruction Work and Materials in Romania: Prequalifications now
2087	$677 million Housing Project in Poland
2088	$120 million Energy Sector Emergency Projects in Estonia, Latvia & Lithuania
2089	$90 million Transportation Project in Bulgaria
2090	Sports Park Development Project in Hungary
2091	Truck Terminal Development Project in Hungary

2092 Large Scale EBRD-funded Construction Project in Hungary; Bids due October 12, 1992
2093 Shopping Center, Duty Free Zone, and Truck Stop Development Projects in Hungary
2094 Mini-USSR Recreation/Theme Park Development Project in Estonia
2095 Electronic Voting Equipment for Czechoslovakia: Bids due August 31, 1992
2096 Airport Development Project in Krakow, Poland
2097 $250 million Road and Bridge Projects in Poland: PREQUALIFY NOW
2098 $55 million to Buy Agriculture, Energy, and Health Products in Lithuania
2099 Cremation Furnaces Purchase and Building Construction in CSFR: Bids due August 31, 1992
2100 EBRD Telecommunications Digitalization Project in Bulgaria
2101 $87.75 million General Procurement Notice: Romanian Transportation Project

Upcoming Trade Events

2510 1992/3 Baltic Trade Fairs, Conferences, and Investment Missions
2535 OPIC Investment Mission to the Baltics: August 23-September 3, 1992
2540 Baltic Trade and Investment Expo, Riga, Latvia: September 23-26, 1992
2545 Plovdiv (Bulgaria) Fair: September 28-October 4, 1992
2546 Free Seminar: The Use of Individual Consultants by the World Bank, Washington, DC: September 20, 1992
2548 Conference: Global Environment Projects: How to Tap Into a Multibillion Dollar Market, Washington, DC, October 5-6, 1992
2550 Bucharest (Romania) International Fair: October 8-15, 1992
2555 UN Investors Forum, Warsaw, Poland: October 26-28, 1992
2560 Energy Efficiency Business Week, Prague, Czechoslovakia: November 30-December 4, 1992

II. Financing Information

1200 Sources for Financing Your Ventures in Eastern Europe
1205 Bulgarian-American Enterprise Fund
1210 Hungarian-American Enterprise Fund
1215 Polish-American Enterprise Fund
1220 Czech and Slovak-American Enterprise Fund
1225 The European Bank for Reconstruction and Development (EBRD)
1230 EBRD Procurement Opportunities Newsletter Ordering Information
1240 USAID Funding for Energy, Environment, and Telecommunications Projects

III. Other U.S. Government Programs in Eastern Europe

1310 U.S. Department of Agriculture Export Programs
1320 U.S. Trade Representative's Activities
2080 $120 million Energy Sector

East European Country Information

Albania
3100 Fact Sheet and Contact Information
3110 Government Contacts
3120 Agricultural Trade Policy and Trade

Bosnia-Hercegovina

3150 Government Officials

Bulgaria

3200 Fact Sheet and Contact Information
3202 Best Prospects for Trade and Investment
3205 Government Contacts
3210 Bulgarian-American Enterprise Fund
3220 Agricultural Trade Policy and Trade

Croatia

3250 Government Officials
3255 Fact Sheet and Background Information

Czechoslovakia

3300 Fact Sheet and Contact Information
3302 Best Prospects for Trade and Investment
3310 Czech and Slovak-American Enterprise Fund
3320 Agricultural Trade Policy and Trade
3370 Finding an Agent or Distributor

Estonia

3400 Estonia Fact Sheet
3410 Useful Business Contacts
3420 Agricultural Trade Policy and Trade

Hungary

3500 Hungary Fact Sheet and Contact Information
3502 Best Prospects for Trade and Investment
3510 Hungarian-American Enterprise Fund
3520 Agricultural Trade Policy and Trade

Latvia

3600 Latvia Fact Sheet and Contact Information
3605 Agriculture Joint Venture Proposals
3610 Useful Business Contacts
3620 Agricultural Trade Policy and Trade
3625 Latvian Fishery Industry Contacts

Lithuania

3700 Lithuanian Fact Sheet and Contact Information
3701 Trade Shows, Investment Missions, and Conferences in Lithuania & the Baltics
3710 Lithuanian Government Contacts
3711 Lithuanian Business Associations
3715 Banking in Lithuania
3720 Goods Available for Barter
3721 Imports to be Bought for Hard Currency
3722 Exports to be Sold
3723 Food Exports to be Sold
3730 Agricultural Trade Policy and Trade
3760 The City of Siauliai Welcomes Foreign Investment
5010 Trade Shows, Investment Missions, and Conferences in Lithuania & the Baltics

Poland

3800	Poland Fact Sheet and Contact Information
3802	Best Prospects for Trade and Investment
3803	American Business Center in Warsaw
3805	Polish-American Enterprise Fund
3810	Useful Government Contacts
3820	Agricultural Trade Policy and Trade
3851	Doing Business in Rzeszow
3852	Doing Business in Plock
3870	Contacts in the Cosmetics Industry

Romania

3900	Romania Fact Sheet and Contact Information
3902	Best Prospects for Trade and Investment
3910	Useful Government Contacts
3920	Agricultural Trade Policy and Trade
3930	Medical Equipment and Supply Industry

Slovenia

3960	Slovenian Independence Creates New Opportunities
3961	Government Officials
3965	Fact Sheet and Background Information

Yugoslavia

3450	Yugoslavia Basic Information
3451	Government Officials
3470	Agricultural Trade Policy and Trade

APPENDIX T - BUSINESS OPPORTUNITIES IN EASTERN EUROPE FOR ENERGY EFFICIENT INDUSTRIAL PRODUCTS

This appendix was made available by the Alliance to Save Energy, 1725 K Street, NW, Suite 509, Washington, DC 20006-1401. It is dated January 1992.

Executive Summary

The economic reform process in Eastern Europe is creating significant potential business opportunities for U.S. firms that market energy-efficient products. Although this process has caused a painful economic readjustment, the Alliance recommends that U.S. firms begin to establish business relationships in the region. By acting now, U.S. businesses will be in a position to profit from the economic expansion that the reforms are likely to produce.

The economic reform process is now well under way in most of Eastern Europe, especially in the energy sector. Energy used in industry is no longer centrally controlled and prices have generally risen toward world levels. Three countries—Hungary, Poland, and Czechoslovakia—have freed most other prices, lowered trade barriers, allowed foreign investment, and made their currency convertible. These reforms are producing a rapid growth in the private sector economy, especially in Hungary.

However, decades of inefficient central planning and subsidized energy prices have resulted in massive energy waste and pollution throughout Eastern Europe. On average these countries use twice as much energy per unit of output as their Western European neighbors. The Alliance estimates that improving the energy efficiency of the region's industrial sector could save $14.5 billion a year in reduced energy costs. We estimate that the potential market for energy-efficient industrial products in the region is $20 billion.

Because industrial facilities throughout the region lack even the most basic energy conservation equipment, there is an enormous potential for low-cost/quick-payback energy saving improvements. The U.S. Agency for International Development's Industrial Energy Efficiency Program recently demonstrated this potential by conducting energy audits and installing U.S. energy-efficient equipment at 48 industrial facilities in Eastern Europe.

The AID program achieved extremely cost-effective energy savings through industry improvement of energy management procedures, use of flue gas and electric demand analyzers, and installation of low-cost flow meters and steam traps. The audits also identified a need for a wide range of energy-efficient products such as lighting systems, energy management controls, high efficiency burners, industrial pipe insulation and energy-efficient motors.

Introduction

Environmental problems and high capital costs associated with energy production are forcing countries throughout the world to examine the least-cost energy path—energy efficiency. From Southeast Asia to Eastern Europe governments and utilities are adopting policies and creating model programs to encourage the installation of energy-efficient technologies. This emerging trend is opening up enormous new markets for U.S. business that manufacture energy-efficient products and provide energy services.

The Alliance to Save Energy's *World Export Program* assists the U.S. business community in identifying and exploring this new market potential. Our program includes seminars, trade missions, and publications. We are supported by restricted project funding from government agencies and unrestricted support from corporate and foundation contributors.

Last year we began our export program with a series of workshops with manufacturers of energy-efficient products to identify industry needs.

During the spring of 1991, the Alliance, supported by the Trade and Development Program and in partnership with the International Institute for Energy Conservation, organized U.S. partnership in a trade mission to Thailand for U.S. manufacturers of energy-efficient products. We also published "A Resource Guide for Exporting Energy Efficient Products" to provide manufacturers with information about government programs that can help them increase exports.

Energy Efficiency in Eastern Europe: An Emerging Policy Agenda Presents Opportunities for U.S. Business

Restructuring the energy sector in Eastern Europe is a critical component of the reform process. On average, these countries use twice as much energy per unit of output as West European countries. Outdated and energy inefficient manufacturing facilities is one of the reasons many enterprises in Eastern Europe produce goods that are not competitive in world markets.

Major environmental pollution problems exist in the region, primarily because of heavy coal use (the use of low-quality brown coal is especially a problem). The close link between energy use and environmental pollution presents Eastern European countries with a difficult problem: how to realize long-term, environmentally compatible growth while addressing short-term economic and social problems. This is a major dilemma, not only for policy-makers throughout the affected region, but also for those in the West who are trying to be of assistance.

Increasingly, policy-makers in Eastern Europe are realizing what their counterparts from around the world already have learned: improving energy efficiency is one of the best ways to meet energy demand in a less capital intensive and polluting way. Adopting policies that encourage the use of energy-efficient technologies can provide the same energy service, but use far less energy (often at one-half to one-tenth the cost of expanding new supplies).

For instance, the United States has reduced energy use per dollar of GNP by 40 percent since the 1970s, in large part by installing more energy-efficient technologies. Western Europe and Japan have also made major improvements in energy efficiency. Given Eastern Europe's high energy intensity, the potential for similar or even greater efficiency improvements is clear.

Eastern Europe represents an enormous new market for energy efficient products of all kinds. Potential products include diagnostic and monitoring equipment, insulation, energy management control systems. high efficiency burners, lighting equipment and controls—in fact, virtually every energy-efficient product currently manufactured.

Demand for these products is emerging as economic reforms cause energy prices to rise. In many cases these technologies are not manufactured or currently available in Eastern Europe, especially in the case of advanced technology. Opportunities also exist for the energy-efficient service industry, which includes engineering, performance contracting, and demand-side management firms. For American businesses that market energy efficient products, the energy problems in the region offer unique new market opportunities for export, licensing and joint ventures. The Alliance estimates that the market for efficient industrial products and services in Eastern Europe is $20 billion.

Energy Supply in Eastern Europe

Energy production in Eastern Europe is dominated by coal, and the pollution stemming from its use is horrendous. Poland produces most of the region's hard coal, (although the quality of Polish coal reserves is declining), but much of its better-quality coal is exported for hard currency. All countries in the region mine and use poor-quality and very polluting soft coal. As a percentage of total energy use, coal use fell by 15 percent between 1971 and 1988, while use of natural gas rose by 60 percent and oil use remained relatively constant. The use of hydro, geothermal and nuclear energy is modest, although nuclear power production grew from virtually nothing a decade ago to 3 percent of total energy use by 1988. In some countries, nuclear power provides a significant portion of electricity generating capacity—26 percent in Hungary, 16 percent in Czechoslovakia, and 15 percent in Bulgaria.

While Eastern Europe has plenty of domestic coal, its proven oil reserved are minimal. In 1989, 88 percent of the region's oil was imported—three-quarters from the former Soviet Union. In addition, one-half of Eastern Europe's natural gas is imported from the former Soviet Union.

Eastern Europe energy policy, while increasingly influenced by the West, will likely also be influenced by the former Soviet Union for many years to come. While the latter likely will remain a major supplier of energy to the region, there is a fundamental difference from before—energy exports are now at market-level prices and payment is in hard currency. It is also likely that the Soviets will encourage Eastern Europe countries to begin gas-for-oil switching, because of Soviet oil production problems and an abundant supply of natural gas.

Energy Demand in Eastern Europe

ENERGY USE AND INTENSITY

During the past four decades, policies in Eastern Europe emphasized increasing economic output at the expense of energy efficiency or market-based energy pricing. Except in Yugoslavia, governments throughout the region regulated and subsidized energy prices. These policies created a situation where energy prices did not cover the cost of production for all energy sources—electricity, gas and oil.

In the past, this lack of energy price signals based on market conditions, in combination with inefficient central planning, has caused serious distortions in these economies. As a result, Eastern European countries are, on average, twice as energy intensive as Organization for Economic Cooperation and Development (OECD) countries, even though per-capita energy consumption is similar to Western Europe. Furthermore, high reliance on coal, especially soft coal, and lack of capital for air pollution abatement equipment has resulted in Eastern European SO^2 concentrations that are four times as high as those in Western Europe.

Fortunately, each of these countries has begun to allow energy prices to rise to world levels or to relate directly to the cost of production.

Eastern Europe's high level of energy intensity is often dismissed by the argument that its industrial sector has a higher proportion of energy-intensive, heavy industry than Western nations. However, as the table above illustrates, only a very small portion of its higher energy intensity can be attributed to industrial composition differences. The primary explanation for Eastern Europe's industrial energy intensity is massive energy waste.

ENERGY EFFICIENCY POTENTIAL

Eastern European energy waste can be reduced by continued reform, use of sound energy management techniques and installation of energy-efficient end-use equipment. If Eastern Europe was as energy efficient as Western Europe its energy intensity would be reduced by 44 percent (.62 to.35 toe/ThUSD). This potential decrease, which takes into account the structural differences between Western and Eastern Europe, means cost-effective energy efficiency improvements could reduce overall Eastern European energy use by 218 million tons of oil equivalent a year. This represents a conservation potential of over 8.65 quadrillion Btu per year, worth $25.8 billion a year in reduced overall energy costs.

More than half of these energy savings could be achieved in the industrial sector (122 Mtoe and

$14.5 billion a year). The Alliance estimates the potential market for energy-efficient industrial products (with paybacks of less than three years) in Eastern Europe is $20 billion. While this estimate is an approximation, it nonetheless demonstrates that enormous potential market exists.

Future Energy Demand

As Eastern European countries continue to reform their economies, energy demand likely will tend to increase for several reasons. Eastern European households presently own fewer home appliances than their neighbors in the West, but increased prosperity should result in more appliance purchases and energy use in the residential sector. Similarly, Eastern Europe's service sector is a much smaller component of economic activity than its counterpart in the West. Energy use in commercial buildings will likely increase as more buildings are constructed and more computers, copiers and other office technologies are purchased.

Additionally, Eastern Europe's transportation sector uses less energy per capita than in Western Europe because the region has fewer autos (although vehicles are often very polluting), and a much higher proportion of people use public transportation than in Western Europe. However, as their economies rebound, the demand in these countries for transport energy, especially imported oil, likely will rise as more and more households can afford to purchase automobiles.

The State of Industrial Energy Management

East European plant managers realized their facilities waste energy, and they are very interested in identifying ways to eliminate waste and improve efficiency. However, they generally lack information about modern energy-saving technologies and management techniques that are common in the West. Under the former system of central allocation of subsidized energy, plant managers had no reason or incentive to learn about—much less install—energy-efficiency improvements. Given concerns about energy-related pollution and the need to pay world prices for energy in hard currency, a powerful incentive clearly has emerged: social and economic survival.

Industrial facilities throughout most of the region lack even the most basic low-cost energy-conservation equipment. For instance, many plants lack flue gas analysis equipment. This technology, which is relatively inexpensive and quickly pays for itself through reduced energy costs, enables technicians to optimize the performance of oil and gas burners, thereby reducing their energy consumption.

Many facilities lack other monitoring and control equipment, such as temperature and flow gauges needed to monitor energy use in equipment, and simple control systems to regulate performance. Fortunately, this type of equipment usually is inexpensive and quickly pays back the investment. Some equipment, especially sophisticated analysis tools, is not manufactured in Eastern Europe and may be one of the first markets for energy-efficiency product exports to the region.

Almost all industrial plants in Eastern Europe lack other energy efficiency measures typically found in the West. These include equipment such as energy-efficient lighting and control systems, high efficiency oil and gas burners and automatic control systems for boilers, energy management control systems for buildings and industrial processes, waste heat recovery systems, and energy-efficient motors and adjustable frequency drives.

PAYING FOR ENERGY EFFICIENCY IMPROVEMENTS

Significant energy savings can be achieved through better energy management and relatively low-cost—and quick payback—conservation improvements. In many instances, savings are so dramatic that enterprises will be able to recover their costs within days, weeks or months.

However, because enterprises have limited hard currency available to them, only measures with a very high rate of return, perhaps 300 percent or more, will be purchased. Although it is hoped that enterprises will be able to pay for these improvements out of their cash flow, additional Western technical and financial assistance also may be needed. In the case of relatively simple technology—steam traps, pipe insulation, etc.—some of these items are locally produced, although these products reportedly are of poor quality.

One of the most critical long-term problems facing government policy-makers and industrial managers in Eastern Europe is overcoming the lack of capital needed to pay for efficiency improvements that have higher first costs and longer payback periods. During Communist rule enterprises were not permitted to make and accumulated profits and most trade was conducted with other Eastern European countries or the former Soviet Union, which paid for goods in nonconvertible currency. As a result, most enterprises started the reform process lacking capital resources, although some now are generating capital. It remains to be seen to what extent they will have the capital needed to invest to efficiency improvements.

The lack of private capital markets in these countries further hinders investment in efficiency. Private banking systems and stock exchanges are only now being established and are modest in size. For instance, only five firms are listed on the Polish stock exchange. This means enterprises face the problem of upgrading their facilities with virtually no internal capital resources or access to equity or debt financing.

Imports of more advanced and, for these countries generally expensive, Western technology will likely be limited to enterprises that export to the West and therefore generate outside hard currency. The exception will be those enterprises able to attract Western business partners, which will provide an infusion of investment capital.

Eastern European countries, unlike the former Soviet Union, belong to the International Monetary Fund and World Bank, although these institutions have been slow to respond with support for energy efficiency investments. Regional energy efficiency could also be stimulated if Western governments and multinational lending institutions adopted policies to provide capital specifically for energy efficiency improvements.

Another potential financing option is private sector-initiated performance contracting, which has been used extensively in the United States as a way to upgrade cash-poor schools and other facilities. U.S. business should explore this opportunity, and government policy should encourage its use.

APPENDIX U - ELECTRIC POWER TRANSMISSION AND DISTRIBUTION SECTORS IN CZCHOSLOVAKIA, HUNGARY AND POLAND

This appendix was made available by the National Electrical Manufacturers Association, 2102 L Street, NW, Suite 300, Washington, DC 20037. It is dated July 1992.

BACKGROUND

This report is the product of a NEMA fact-finding mission conducted as part of a U.S. Trade and Development Program grant focused on the electric power transmission and distribution sectors in Czechoslovakia, Hungary and Poland.

The NEMA mission took place June 22-29, 1991, with on-site visits in Czechoslovakia (Prague and Bratislava), Hungary (Budapest), and Poland (Warsaw). The listing of delegation participants and individuals visited in each country are contained in the annex to this report.

NEMA expresses its appreciation to the U.S. Foreign Commercial Service officials and their staff in Czechoslovakia, Poland, and Hungary. Without their assistance and support, this report would not be possible.

Prior to the fact-finding trip, the delegation reviewed market data and country information as prepared and supplied by the U.S. Department of Commerce. A selected listing of those materials are also shown in the annex to this report.

This report is one aspect of a NEMA/TDP initiative aimed at assessing business opportunities for American firms supplying transmission and distribution equipment to each of the three eastern European countries. For more information please contact Mr. Kyle Pitsor, NEMA manager of international affairs.

CZECHOSLOVAKIA

The Federation of Czechoslovakia is composed of a Czech Republic and a Slovak Republic. Each have their own utility entities. Up to 1990, generation, transmission and distribution functions were the responsibility of the Czech Energy Works and the Slovak Power Enterprises. In 1990, distribution was separated from generation and transmission. There are now eight distribution companies in the Czech Republic and three in the Slovak Republic. The generation and transmission grids in the two republics are separate. At present, all entities are state owned.

THE CZECH REPUBLIC

The Delegation met with representatives from the Czech Energy Works and the Ministry for Economic Policy and Development of the Czech Republic. A session was also held with the director of the energy department in the Federal Ministry of Economy.

Transmission

Czech Energy Works operates 7,000 kilometers of 440 kV and 220 kV lines as well as some 100 kV (considered distribution). Most of the equipment is old and needs to be replaced. There are fault current problems and a considerable amount of maintenance is required. Most of the installed equipment is either from local manufacturers or from the USSR. As the 220 kV lines need replacing, they will covert to primarily a 440-to-110 kV system. They would like to connect to the Western European transmission grid (UCPTE) by 1995. This will require AC-DC-AC interconnect and converter equipment. They already have an AC tie with the USSR and Hungary which they would like to discontinue.

Distribution

The Czech Republic has eight companies responsible for distribution, 110 kV down through 380 volts. Primary distribution voltages are 10 kV and 22 kV, three phase. The 380 volt operations are located in rural areas and are mostly overhead lines. In the larger cities, they use 22 kV to 380 volt or to 220 volt. One district has some 110 volt. Prague is a combination of both 220 and 110 volt. They would like to convert all of Prague to 35 kV to 110 kV by 1995. Installed equipment is old and needs to be replaced. There is a considerable amount of work needed on both the rural lines as well as the urban systems. A key issue will be financing since the national grid is in bad repair and will get attention first.

Economics

The price of electricity to industrial users was raised 100 percent last year and the residential prices were raised 80 percent this year. Further raises are necessary as they desire to get pricing to comparable West European levels by the end of 1992. The present pricing structure does not cover cost to deliver from generation to customer. Inflation is running nearly 30 percent for 1991, and another devaluation of the currency is expected. The Government needs to build reserves so as to be able to apply for credits/loans and is negotiating with the World Bank for a $250 million loan that will encompass the interconnection project with the Western European transmission grid (UCPTE).

Energy privatization plans are under development. The Czech Ministry currently envisages all entities to be stock companies beginning January 1, 1992, with the State having 100 percent of the stock. By June 1, 1992, the State's share will drop to 60 percent, with the 40 percent having been offered to foreign firms and private investors. While there may be no limitation of foreign ownership in some sectors of the economy, in energy the State will maintain a 60 percent holding. This scheme is still under development and may change. A British firm is advising the ministry officials on energy plans and restructuring of the energy industry. Reaction from the entities is that the transition is too fast.

Equipment

Much of the equipment is very old (30 to 40 years) and will need to be replaced. Present equipment does not comply with IEC standards but they plan to comply with IEC standards in purchasing new equipment. Most equipment now in service is from local manufacturers (Skoda, Fucik) or from former East European country suppliers. Oil and air insulated circuit breakers (110 kV) are from local firms. They locally manufacture some SF6 circuit breakers and switches but have purchased some from German transformers are from the USSR and Yugoslavia. Surge arresters are produced by Polish, Soviet, or (East) German manufacturers. Metal oxide arresters are (West) German. Power transformers are mostly from local firms with some (110 kV) from Yugoslavia, some 220 kV transformers have been purchased from (East) German suppliers, and some (440 kV) from British and Soviet firms. Many of the 440 kV transformers need to be replaced; they are 20-30 years old and have very high losses. Underground 110 kV cable is 80 percent PE crosslinked and 20 percent oil pressure. There is about 56,000 kilometers of 10 kV and 22 kV underground PE cable. Local manufacturing firms would like to form joint ventures with technology transfer as part of the joint venture arrangement.

THE SLOVAK REPUBLIC

The delegation met with representatives with the Slovak Power Enterprises, the West Slovak Distribution Company, and with an official of the Slovak Ministry of Economy.

Transmission

The Slovak Power Enterprises (SEP) operates the generation and transmission network in Slovakia. SEP sells and imports electricity directly to the distribution companies, and also directly to certain industrial customers. The system operates at 400-to-220-to-110 kV and at 400-to-110 kV. They want to eventually operate only a 400-to-110 kV system. Given the mountainous terrain, the grid is characterized by multiple systems, compact linens and 350-450 square millimeter conductor lines. There is under study the possibility of DC transmission. SEP plans to complete its survey in the near future on network needs in order to participate in the negotiations for the interconnection project to UCPTE. The list may include extensive equipment replacement needs and enhanced system protection and automation requirements.

Distribution

There are three distribution companies—West Slovakia Electricity Works, Central Slovakia Electricity Works, and East Slovakia Electricity Works. Electricity dispatching is managed by SEP and the distribution companies purchase electricity at 110 kV from SEP. The three companies, in turn, sell the electricity at 110 to 30 county entities (10 counties per distribution company). The distribution companies have a combined 110 kV line of 4.290 kilometers; SEP has 815 km of 220 kV and 1,509 km of 400 kV. In rural areas the cables are above ground and in the city they are underground. The distribution firms have independent purchasing authority, but use SEP assistance because it has more experience.

Economics

Electricity prices for industrial customers increased 80 percent on April 1, and residential prices will increase 80 percent on September 1, 1991. As in the case of the Czech Republic, the Slovak Republic is also planning to privatize the state-owned enterprises. The list of companies to be privatized is being developed and when completed the Slovak Government will decide which entities will be offered as joint stock companies. Some companies will be offered a first phase and other companies will be offered in a second phase. The fate of the energy entities is not yet decided.

Equipment

As in the case of Czech Energy Works, the equipment in Slovakia is older, generally locally manufactured, and in need of upgrading and replacement. For instance, Skoda produces 110 to 220 kV distribution transformers, air blast switches, and SF6 breakers. Fucik, another local manufacturer, supplies SF6 breakers. Fucik, another local manufacturer, supplies SF6 switches (110 to 220 kV).Circuit breakers are oil and switches are air blast. They have been supplied by local firms and German ones. In most cases, transformers are oil-cooled, but also have air-cooled network transformers. They have had some problems with the paraffin oil leaking and causing ecological problems. They have purchased a few Belgium and British transformers. While most cable is local, they have purchased some underground PE crosslinked cable from France. Slovak Power Enterprises is applying for Republic Ministry trade credits for equipment purchases.

HUNGARY

The Hungarian Ministry of Industry and Trade (MIT) exercises state control and general supervision over the electric energy industry. The Hungarian Electricity Board (MVMT) is responsible for the transfer and distribution sale of electricity. MVMT is organized in the form of a trust and supervised by MIT> The production of electric energy takes place primarily in the power stations belonging to MVMT. Several companies belong to the trust, including 11 power stations, 1 company managing the high-voltage transmission network (OVIT), 6 distribution companies, and 4 each attending to maintenance, construction, electrical equipment manufacturing, and investment.

The delegation met with representatives of the Hungarian Electricity Board (MVMT), the National High-Voltage Transmission Line Company (OVIT), Budapest Electric Works (ELMU, the distribution company serving Budapest), and Power Station Investment Company (ERBE, the power plant construction firm).

Transmission

The National High-Voltage Transmission Line Company (OVIT) has responsibility for the construction, operation and maintenance of the Hungarian basic network on 120, 220, 400 and 750 kV. It operates 268 km of 750 kV lines (connected to the USSR grid), 1600 km of 400 kV lines, 1300 km of 220 kV linens, 2,500 km of 120 kV lines (maintenance for the 6 distribution companies), and 37 transformer substations with 13,000 MVA of capacity. OVIT has the task of the development of the grid from engineering to construction. It has about 700 employees including 200 engineers and skilled workers. The future activities of OVIT will be largely determined by the Hungarian strategic energy plan which will involve the building of 400 kV network lines, direct converters with at least 1100 MVA capacity, modernization of the telecommunications and automation of the Hungarian network. A German firm has been retained to study Hungary's substation dispatching system study and a U.S. firm is studying the SCADA/network control. In consultation with MVMT, OVIT and ERBE (Power Plant Investment Company) will be making recommendations on power plant and interconnection projects. ERBE has 400 employees, most of whom are engineers. ERBE plays a key role in managing construction investments and reconstruction needs of power stations as assigned by MVMT or its companies. ERBE attends to tenders, selection of equipment, and bid evaluation of equipment purchased through World Bank loans.

Distribution

Electricity is purchased by the six regional distribution companies at 120 kV. In the country, distribution primarily occurs on 20 kV lines while in the cities (over 100,000 in population) 10 kV lines are more common.

ELMU, the Budapest Electric Works, has responsibility for Budapest and its surrounding area. Budapest has about 2 million inhabitants, which consumed about 8,200 GWh in 1989 (2,622 GWh were household). ELMU purchases 90 percent from the power grid, with the remaining 10 percent from local steam plants in the district. ELMU distributed at both 20 kV and 10 kV, with all secondary power at 380 volts, three-phase. Most is underground, with a small amount overhead at 20 kV. It operates 120 k of 35 kV open air lines, 3,100 km of 20 kV open air lines, 170 km of 30 kV underground cable, 220 km of 20 kV underground cable, 4300 km of 10 kV underground cable.

Economics

Electricity prices are some 30 to 50 percent below comparable Western European prices. Residential prices were raised 50 percent last year. The Government will likely increase industrial prices in early 1992. Overall demand has been flat the last several years, with industrial demand declining (6 to 9 percent). Residential demand is up, about 6 to 7 percent a year. Total growth is expected in the 2 to 3 percent range. (Nearly 30 percent of total electricity is imported from the USSR. The goal for 1991 is to limit USSR imports to 20 percent or below.) Studies are underway to connect into the West European grid, UCPTE.

Privatization plans are under development. Beginning January 1, 1992, the energy entities will be transformed into stock companies. For the power generation companies, the State will retain

100 percent of the stock. For the distribution companies, the State will retain 51 percent of the stock. Each new company will make their own decisions on equipment purchases, with the MVMT setting standards and conducting evaluations. For ERBE, the State will have 70 percent of the stock and the employees will have the remaining 30 percent. With respect to Hungarian electrical manufacturers, there are already a number of joint ventures with Western European equipment firms including German, Swiss/Swedish, and French. Foreign firms can hold up to 51 percent of the equity in these joint ventures. While there are no currency barriers in purchasing imported equipment, there is little hard currency available; financing of projects/equipment is necessary.

Equipment

Most of the equipment currently in service was produced by local Hungarian firms. The equipment is generally old and in need of replacement and upgrading. Hungary has an extensive project underway to convert to IEC standards (or CENELEC standards) with the need to deviate in some areas.

In the transmission area, the interconnection with the UCPTE will require AC-DC-AC conversion equipment and regulation devices. Power cable is produced domestically, and is PE crosslinked. Transformers are domestic although an Italian firm has entered into a joint venture with a Hungarian manufacturer. Some SF6 circuit breakers have been purchased from a Swiss/ Swedish firm. They also have 10 kV and 20 kV mineral oil circuit breakers. In the distribution area, the operation of the system is expensive and suffers from overload problems. Plans are underway at ELMU to construct a new sub-station with a 25 MVA transformer and 10 kV and 120 kV SF6 switchgear and vacuum interruption equipment. Cable is crosslinked polymer and supplied from local and foreign firms. They have had some splicing problems with the foreign cable. Insulators are all porcelain and are produced domestically. They have no metal oxide surge arresters, and have had explosion problems in the 20 kV areas. They would like to implement improvements in substation automation.

A feasibility study, under contract with a U.S. firm, is under way regarding load management of transmission/distribution. The World Bank is likely to finance a load management loan based on the feasibility study. Also, there has been some interest in exploring demand side management with a Canadian entity.

POLAND

In 1990, the Union of Electric Power Industry (WEWB) was dissolved and a new Polish generation, transmission and distribution scheme was formed. There is one national transmission grid company, the Polish Power Grid Company and 33 distribution companies of which 32 have joined to form the Association of Distribution Companies (the Warsaw Distribution Company is not a member). There are 28 power generating companies represented by the Power Utility Industry. All are state owned. The Ministry of Industry supervises, sets regulatory policy and creates the legal framework for the energy sector.

The delegation met with representatives of the Polish Power Grid Company, Poznan Distribution Company (representing the Association of Distribution Companies), and the Ministry of Industry.

Transmission

The Polish Power Grid Company (PSE) operates the central dispatching office and 5 regional offices. PSE is responsible for purchasing electricity from the generating entities, dispatching and selling it to the distribution companies, and planning for system procurement needs. The transmission system is composed of 100 km of 750 kV lines built in 1984 and connected to the USSR grid, 4,000 km of 400 kV built in the 1970s and 1980s serving as the main transmission basis, and

8,000 km of 220 kV lines that are older (1950s and 1960s), somewhat unreliable, and serve only local needs. PSE plans only to maintain, not extend, use of 220 kV lines. They use 2-1100 MVA autotransformers at the 750/400 kV switching station.

PSE has underdevelopment plans to extend the 400 kV grid in the northeast, add some 6 new sub-stations, and connect to the UCPTE system. The interconnection is 5 to 6 years away and will not need DC converters. The grid is very reliable, but there exists a lot of reactive power, over voltage problems on certain downstream equipment (on 400 kV) and some transmission line losses. There are no major peak needs (winter is peak period), so transmission projects are limited to reconstruction of existing system and thermal upgrading of lines and sub-stations. There are short circuit faults on 220 kV network lines which exceed nominal resistance by 15 to 25 percent, for both single and three phase.

Distribution

The Polish distribution system is composed of 33 companies, legally independent from each other since January 1990. They purchase electricity only from PSE and sell it to the end-user. The prices are presently set by the Ministry of Finance, although this is to change.

In 1992, the Federal government's subsidies to the coal industry (on which over 90 percent of Polish power is generated) will end. This will allow the mining entities to negotiate with the power generators the cost of purchased coal. By the end of 1992, in turn, the generators will negotiate with the transmission company (PSE) the cost of generated electricity. The Ministry of Finance, however, will set the maximum price PSE can sell electricity to the distribution companies. The distribution companies will sell the electricity to residential and industrial consumers at prices they determine; however, the Ministry of Finance may still regulate maximum prices.

There is recognition of the need to reorganize the distribution companies and reduce the number of firms. A coordinating association representing 32 distribution firms has been organized. While each company has its own purchasing policies, the firms are exchanging information on products and purchasing decisions. Some 70 percent of the distribution equipment is in urgent need of replacement due to age and inability to remotely control the equipment. Only 50 percent of the 110 kV mid-voltage substations have remote telecontrol equipment and software. Distribution voltages are 110 kV, 15 kV (85 percent), 20 kV (new) and 380 volts; all voltages are three phase. The distribution system is not grounded; they use a floating neutral or neutral resistor.

Economics

Major changes in the energy sector are focussed in decentralization, privatization, and self-financing. Privatization of the generation and transmission entities will occur by first creating State stock companies and then offering some stock for private purchase. Plans also call for the merging of the 33 distribution companies into 15 firms that will also become stock companies.

Electricity is still under-priced in terms of reflecting the real cost of generation and distribution. Residential prices were raised 100 percent and industrial prices went up 125 percent in May 1991. It is anticipated that industrial prices will be adjusted upward later this year by another 10 to 15 percent. It was stated that residential prices may be 40 to 70 percent higher than industrial prices.

Equipment

In both the transmission and distribution areas, equipment has been supplied by a variety of Polish (50 percent) and East European countries. They have equipment needs in all areas of both transmission and distribution, from both transformers to breakers, high to low voltages. The Polish Standards Institute is currently evaluating using IEC standards and harmonizing current product standards to IEC standards. A product certification scheme is being considered by the Polish Government.

Some 15 types of air-disconnect switches are being used purchased from Polish, Yugoslavian,

Soviet and French suppliers. Some SF6 circuit breakers are on the 110 kV system, and their use will be expanded to the 220 and 400 kV system. A Polish circuit breaker manufacturer has a licensing agreement with a Swiss/Swedish firm. At the 15 kV level, load break switches are being quickly introduced to replace air-insulated ones. Arresters are mostly Polish (no metal oxide arresters). Insulators are porcelain (Polish-East German) and a small number are glass (USSR). The 15 and 20 kV underground cable used in cities is 60 percent paper-insulated, and 40 percent PVC crosslinked. There are some 200,000 medium and low voltage substations, all in above ground concrete structures. Transformers are oil only (dry-type are too noisy), not grounded but with neutral in cable system. There are protection concerns in the villages due to lack of grounding. Urgent need for SF6 breakers, load interrupters, solid state breakers, and remote control equipment.

APPENDIX V - TRIP REPORT ENERGY ENGINEERS IN EASTERN EUROPE

This appendix was authored by Prof. W.H. Mashburn, Mechanical Engineering Department, Virginia Tech. It is a trip report for a group of members of the Association of Energy Engineers who toured Eastern Europe in 1991. It was provided for this report by the Association of Energy Engineers.

I spent the last week of June and first week of July touring Eastern Europe as part of a program set up by the Association of Energy Engineers. The purpose of the tour was to meet with key officials in each of the EU countries to learn more about their energy and environmental situation, and how we might form linkages to assist with their problems. My main objective was to obtain more information in support of an upcoming RFP from U.S. AID to do comprehensive training in Eastern Europe in energy and environmental subjects.

We met with officials in East Germany, Poland, Hungary, and Czechoslovakia. There are several problems that are common to all. These are:

1) The number one priority of each country seems to be privatization. Each has some type of trust set up to move the process along, especially in the power generation business, which will end being a mixture of private and public ownership.
2) Most internal power generation is done with low quality coal such as lignite as the major fuel. This is a major cause of environmental degradation in those countries in that both the mining and burning contribute to pollution. They are interested in both clean coal technology and the purchase of a higher grade of coal.
3) They predict a decline in the use of electric energy until after the year 2000 because of the decreased production in their industrial sector. Very simply, because of poor quality, there is a limited market for their products now that the Russian market has declined.
4) They are very concerned about the environmental conditions in their countries, and all have some type of ministry of environmental control. Environmental regulations have been established that they and all other European countries have subscribed to, and in many cases they are more stringent than those in this country.
5) Energy prices have increased in the past year. In some cases, they have quadrupled. This is due to the fact that their governments have removed subsidies, outside purchases—such as from Russia—require hard currency, and barter is no longer accepted.
6) All have high hopes for a more competitive source of natural gas when a line from Norway is completed.
7) There are other related problems such as the mind set and work ethnic of people who have for the past forty five years labored under a system that did not require or even desire people to have initiative.

Some specifics and contacts for each of the countries visited are given below:

EAST GERMANY

East Germany, which is now really just Germany, is looking to West Germany for technical and financial assistance. In fact, they envision getting to the point where they can assist the other EU countries because of their unique position and experience.

Most of their electricity is generated with lignite. They are interested in new clean technology to continue to burn it. They have two nuclear reactors. Four Russian designs were under construction, but have been stopped. They recognize a weakness in the controls and safety of Russian designs. Industry is now requiring only 70% of the original power requirements because of decline in production. They have ample capacity and do not import electric power.

They are concerned about unemployment of many of the power industry employees. Many of the engineers will be retrained and moved to environmental issues.

Economic concerns are:
Future economic growth and employment
Stability of currency
Foreign relations
Common European Market by 1992
East German production decreased 40% since 1989
Unemployment rate is increasing

Energy Issues are:
Need to develop a modern energy system
A concept for modernization is due by the end of 1991
They recognize the importance of such a plan for potential investors
Domestic energy sources are promoted Very low efficiency is a problem

Some reasons given for poor energy efficiency were:
False economic structure—subsidized costs
Service industry poorly trained
Strong tendency toward high energy intensity industries
Poor connections—high line losses
Inefficient housing
30% of boilers are over 30 years old
Inefficient processes
Poor end usage

Contacts in East Germany:

Hans-Jurgen Petrick
Dipl.Geophysiker
Bundesministerium fur Wirtschaft
Aufsenstelle Berlin
Unter den Linden 44-60
0-1080 Berlin
Tel. (Ost) 23-35777
(West) (030) 39985-351

Eckhard Netzmann
Kraftwerks-und Anlagenbau Ag Berlin-Marzahn
Vorstandsvorsitzender
Allee der Kosmonauten 32
0-1140 Berlin
Tel. 5 46 50

Kurt Beckervordersandforth
Mitglied des Vorstandes
Kraftwerks-und Analagenbau Ag Berlin-Marzahn
Allee der Kosmonauten 32
0-1140 Berlin
Tel. 5 46 50

POLAND

The privatization process will divide the electric utility industry into three segments:
Generation—which will remain state owned
Transmission—state and stock companies
Distribution—eventually will be stock companies

Purchase of coal from outside sources will be competitive because of the low efficiency of production of their own industry.

The Ministry of Environmental Protection has strict environmental laws that will go into effect in 1993-4. Money from fines can be used by offender to construct pollution control equipment. 25% of their power stations are over 25 years old, so there is a problem with modernization. They have money from the World Bank to update central heating systems. Westinghouse has completed a study of five power plants. They plan to update the power plants and will include environmental controls.

They plan to have surplus power to sell, and are considering and underseas line to Sweden. Sweden, in a surge of purity a couple of years ago, voted to remove all nuclear plants, which now supply approximately 30% of their total electric power. A conversation with a Swede I met in Vienna indicated that reality was now beginning to overcome their over reaction.

They have high hopes for the gas line from Norway. A new rate structure for electrical energy is being developed to be in line with free market pricing of energy. The average cost per kw-hr is 5 cents.

Contacts in Poland:

Roman Luczkiewicz M.Sc(Eng.)
Adviser to the Minister of Industry
Minister of Industry
4 Wspolna, 5th fl. room 5106
00-926 Warsaw, Poland
phone: 29 57 76

Mgr inz. Andrzej Strzelecke
Z-ca Sekretaraza Generalnego
Stowarzyszenie
Elektroykow Polskich ul. Czackiego 3/5
00-043 Warszawa
Tel. 27-38-79

Dr. Jan Solinske
Secretary of the Polish National Committee
of the World Energy Conference
Power Research Institute
8 Mory Str.
01-330 Warszawa Poland
Tel. 693-11-13

Mgr inz. Janusz Raszewske
Sekretarz Generalny
Stowarzyszenia Elektrykow Polskich
ul. Czackiego 3/5
00-04 3 Warszawa
Tel. 26-86-01

Mgr inz. Jacek Szpotanske
Prezes, Stowarzyszenia Elektrykow Polskish
ul. Czackiego 3/5
00-04 3 Warszawa
Tel. 26-85-76

Bronislaw Opacki Jr
Obozowa 91 m 4
01-443 Warszawa
Tel. 36-41-41

HUNGARY

Hungary is heavily involved in the privatization process with approximately 10% of its industry now in private hands. 50% of its electric power is generated with nuclear energy. 64% of its total energy is imported mainly from Russia in the form of oil, gas, and some electricity. Hard currency is required for payment. This situation makes energy conservation critical. Their situation with regard to production, pollution and incentives is pretty much the same as the other EU countries.

Contacts in Hungary:

Morenth Andras
Head of Department
Ministry of Industry and Trade
Directorate-General for Energy Policy
Department of Energy Cooperation
1525 Budapest, Pf96
Tel. (36-1)155-83-72

Dr. Techn. T. Jaszay C.HASc.
Assoc. Prof. Deputy Director
Institute for-Thermal Energy Systems Engineering
Technical University of Budapest
Budafoke ut 4 Budapest H-III
Tel. (36-1)166 68 08

Tamas Zettner
President Professor
Hungarian Scientific Society of Energy
Budapest, P.O.B. 451
H-1372
Tel. 36 1 153 2751

Gyorgy Hatvani
President
Ungarischer Elektrotechnisher Verein
H-1372 Budapest Pf. 451
Budapest V., Kossuth Lajos ter 6-8
Tel. 36 1 153 0117

CZECHOSLOVAKIA

Power is generated with 30% nuclear and 70% coal. Hard coal comes from Bohemia and brown coal (lignite) from Slovakia which is a province of the country. Industrial production is down 20% from previous years and is projected to decline until the year 2005. In order to improve environmental conditions, they plan to convert much of the coal to natural gas most of which will be purchased. They also plan to have more nuclear power with plants purchased from Russia. They are negotiating with Westinghouse for controls and safety features. Austria is strongly opposing this plan for nuclear power.

District heating is provided to 30-40% of the buildings. Private homes are heated by gas or coal. Energy costs have quadrupled since May.

They have a building conservation program, but have found the structures built by the communists to be very inefficient. Agencies and universities have started some training in energy. There is a lack of instrumentation, materials, and manpower. Cheap energy for the last 40 years has made it difficult to overcome the economic impact of a free market.

Contacts in Czechoslovakia

Ing.Vaclav Simunek
Director of Department
Federal Ministry of Economy
Pravada
Tel. 389 21 90

JUDr. Jan Krajc
Slovak Government Agency
for Foreign Investment
Pravada
Tel. 389 26 43

Additional Information

From my work this spring in Romania, I have additional information on this country which includes and industry profile report and industrial audit reports.

The U.S. Chamber of Commerce has a datafile report of each of the eastern European countries. These reports can be obtained from them at the following address:

Central and Eastern European Trade and Technical Assistance Center
International Division
U.S. Chamber of Commerce
1615 H. Street, N.W.
Washington, DC 20062

APPENDIX W - USDA REPORT PROGRAMS FOR EASTERN EUROPE

This appendix was made available by the U.S. Department of Commerce.

USDA EXPORT PROGRAMS FOR EASTERN EUROPE

The U.S. Department of Agriculture's (USDA) Foreign Agricultural Service (FAS) offers several important programs that could improve the outlook for U.S. agricultural exports in Eastern Europe. These programs, which include credit guarantees, Export Enhancement Program (EEP) initiatives and targeted promotional assistance, can help build potential U.S. agricultural markets in Eastern Europe's new era of democratization and privatization.

Credit Guarantees

FAS operated two export credit guarantee programs on behalf of the Commodity Credit Corporation (CCC) designed to increase the willingness of the private banking system to extend credit for agricultural exporters—U.S. banks or other financial institutions—in case a foreign buyer's bank breaks its promise to pay. Up to now, Bulgaria, Czechoslovakia, Hungary, Romania and Yugoslavia have all qualified for credit guarantee programs.

The primary difference between the two CCC credit guarantee programs is the length of credit periods they cover. The GSM-102 pertains to short-term loans of six months to three years, and the GSM-103 covers medium-term loans for three to 10 years. Since 1982, Yugoslavia and Hungary have purchased approximately $450 million in U.S. agricultural commodities under GSM-102. For fiscal year 1992, a $50 million GSM-102 program for protein meals, oilseeds and cotton has been established for Romania and a $10 million program for cotton has been established for Hungary. In addition, another GSM-102 program allocating $20 million worth of wheat and soybeans to Bulgaria was recently proposed.

Due to its medium nature, the FAS predicts that GSM-103 programs could become more prominent in the future, especially now that the International Monetary Fund is requiring heavily indebted Eastern European countries to reduce short-term obligations before it formulated financial aid packages.

For more information on how to request guarantee allocations, please feel free to contact the Credit Commodity Corporation, Operations Division, U.S. Department of Agriculture, Washington, DC 20250-1000; telephone: (202) 720-3224.

Other Programs

The USDA's Export Enhancement Program (EEP) enables U.S. exporters to meet prevailing world prices for targeted commodities and destinations, including countries in Eastern Europe. This program is designed to combat and discourage non-competitive practices.

The EEP helps U.S. exporters meet prevailing world prices by making bonuses available to them. The program was not established to undercut world market prices, but to help agricultural products produced by U.S. farmers meet competition from subsidizing countries, especially the European Community.

EEP bonuses can be requested by anyone having legitimate concerns about subsidized competition in a specific market. For more information on how to qualify for participation in the program, please call the CCC Operations Division at (202) 720-5540.

Another program administered by FAS is the Market Promotion Program (MPP). Mpp helps those U.S. exporters disadvantaged by unfair practices finance promotional activities for agricultural products using surplus stocks or funds from the CCC. USDA has approved MPP programs to promote a wide variety of commodities throughout Eastern Europe. Activities financed by the programs vary from commodity to commodity, and include activities such as market research, practical demonstrations for food professionals and point-of-sale contact with consumers.

For more information on the MPP program and how to submit a proposal, please call the MPP Marketing Operations Staff at (202) 720-2705.

AGRICULTURE OPENS DOORS IN EASTERN EUROPE

As the Eastern European markets open to trade with the West, the agriculture sector is at the forefront of targeted export markets. Whether your firm is new to exporting or is experienced in foreign trade, the Department of Agriculture's Foreign Agriculture Service (FAS) can provide invaluable export assistance.

In addition to export information and counseling aimed principally at new agriculture exporters, the Foreign Agricultural Service provides the following trade assistance information:

Agriculture Trade Assistance Folder

Helps new and experienced exporters become familiar with the wide variety of trade assistance programs available through FAS and other Government agencies. Contains samples of FAS publications such as "U.S. Legislation Counters Unfair Foreign Trade Practices" and explains how the publications may be ordered. FAS may also aid exporters in contacting the proper Federal offices administering the various trade remedy laws.

AgExport Action Kit

Helps put U.S. suppliers of food and agricultural products in touch with foreign importers.

Contains order forms for listings which identify overseas buyers of specific products as well as information on agricultural trade shows.

Trade Policies and Market Opportunities for U.S. Farm Products (Annual Report). Identifies the policies used by other governments to help their agricultural exporters compete against U.S. exporters and identifies import barriers to U.S. farm products.

The AgExport Action Kit and Agriculture Trade Assistance Folder can be ordered from:

Trade Assistance and Planning Office
Foreign Agriculture Service
U.S. Department of Agriculture
3101 Park Center Drive
Suite 1103
Alexandria, VA 22302
Contact Person: Jim Ross, Director
(703) 756-6001

The FAS's Export Credit Guarantee Program covers the risk of default by the purchaser's bank. For further information, please contact:

U.S. Department of Agriculture
Commodity Credit Corporation (CCC) Operations
Room 4513-South Building
Washington, DC 20250

Contact Person: Richard Godsey, Branch Chief
(202) 720-3224

Export Credit Program Development Division
Room 4524-South Building

The Eastern Europe and Soviet Secretariet within the Department of Agriculture also acts as a contact point for private agricultural and agribusiness organizations and individuals interested in Eastern Europe. This office may be contacted at:

U.S. Department of Agriculture
Foreign Agriculture Service
Contact Person: Alexander Gilchrist, Coordinator
Room 6506-South Building
(202) 720-0368

It should be noted that this office will not assume any programmitic responsibilities, but will assist in the overall coordination and monitoring of export programs to Eastern Europe.

It will greatly benefit any person interested in exporting agricultural goods to Eastern Europe to educate his or herself as to the range of the FAS's services provided to new and veteran exporters.

APPENDIX X - SOURCES FOR FINANCING VENTURES IN EASTERN EUROPE

This appendix was made available by the U.S. Department of Agriculture, Commodity Credit Corporation (CCC) Operations, Room 4513-South Building, Washington, DC 20250.

TRADE FINANCING

Export-Import Bank of the United States (Eximbank)

The Export-Import Bank of the United States (Eximbank), an independent U.S. Government agency, supports the export of U.S. goods and services through a variety of loan, guarantee, and insurance programs. The types of programs and monetary limitations available for any given country depend upon Eximbank's periodic country risk assessments. Bulgaria, Czechoslovakia, Estonia, Hungary, Latvia, Lithuania, and Poland are eligible for Eximbank's assistance.

Direct and Intermediary Loan Programs: Direct loans are available to Eastern European buyers of U.S. goods and services. Intermediary loans fund responsible parties that extend loans to Eastern European buyers of U.S. capital goods and services. Medium-term (1 to 5 years) and long-term (more than 5 years) loans and guarantees are available for Czechoslovakia, Hungary and Poland.

Working Capital Guarantee Program: Eximbank provides repayment guarantees on secured, short-term loans of pre-export, working capital to qualified small and medium-sized exporters. Pre-export activities include inventory, raw materials, marketing and manufacturing.

Commercial Bank Guarantee Program: Eximbank may provide repayment protection for commercial bank loans to creditworthy buyers of U.S. capital goods and services in eligible countries. These guarantees may be combined with an intermediary loan.

For more information, please contact: Marketing Division, Export-Import Bank of the United States, 811 Vermont Avenue, NW, Washington, DC 20571; telephone: (202) 566-4490; fax: (202) 566-7524.

Companies in Arizona, California, Idaho, Nevada, Oregon and Washington may contact Eximbank's West Coast Office for program information, counseling and training support: Export-Import Bank of the United States, Suite 9103, 11000 Wilshire Blvd., Los Angeles, CA 90024; telephone: (310) 575-7425; fax: (310) 575-7428.

Foreign Credit Insurance Association (FCIA)

The FCIA, an agent of the Eximbank, is a private association that provides export credit insurance to cover commercial and political losses. The FCIA facilitates exports by (1) protecting against the failure of foreign buyers to pay for commercial or political reasons; (2) encouraging exporters to offer competitive terms of payment; (3) supporting an exporter's prudent penetration of higher risk foreign markets; and (4) providing exporters and their banks greater financial flexibility in handling overseas receivables. FCIA thus lowers costs, making American exports more competitive. Service and leasing contracts may also be covered.

FCIA offers short-term single buyer, multi-buyer export, medium-term, new-to-export, and umbrella policies. Bulgaria, Czechoslovakia, Estonia, Hungary, Latvia, Lithuania, Poland, and Romania are eligible. FCIA's country programs and eligibility are based on Eximbank's country risk assessments. For more information, contact the FCIA Regional Office nearest you:

1. New York: 40 Rector Street, Suite 1622, New York, NY 10006
Phone: (212) 227-7020; Fax: (212) 513-4704
2. Chicago: 19 S. La Salle Street, Suite 902, Chicago, IL 60603
Phone: (312) 641-1915; Fax: (312) 641-2292
3. Miami: 80 SW 8th Street, Suite 1800, Miami, FL 33130
Phone: (305) 372-8540; Fax: (305) 372-5114
4. Houston: 222 N. Sepulveda Blvd., Suite 1515, El Segunda, CA 90245
Phone: (310) 322-1152; Fax: (310) 322-2041

U.S. Department of Agriculture

The U.S. Department of Agriculture supports the sales of American agricultural products (including fresh water fish and forestry products) through the Commodity Credit Corporation (CCC) and the Export Enhancement Program (EEP). The CCC guarantees repayment of short-term loans (6 months to 3 years) made to eligible countries purchasing American farm products. Bulgaria, Czechoslovakia, Estonia, Hungary, Latvia, Lithuania, and Romania are eligible for this program. The EEP challenges unfair trade practices by our major competitors.

Contact: Export Credit Program Development Division, U.S. Department of Agriculture, Room 4524, South Building, Washington, DC 20250; Phone: (202) 720-5319; Fax: (202) 690-0051.

INVESTMENT ASSISTANCE

Overseas Private Investment Corporation (OPIC)

The Overseas Private Investment Corporation, a self-sustaining, independent agency of the U.S. Government, promotes economic growth in developing countries by encouraging U.S. private investment. OPIC (1) assists in financing investments through direct loans and/or loan guarantees; (2) insures investments against a broad range of political risks; and (3) operates a match-making service. OPIC insurance covers a variety of risks, such as expropriation; currency inconvertibility; and losses due to political violence, war, and civil disturbances.

OPIC's all-risk loan guarantees, issued to U.S. lending institutions, typically range from $2 million to $25 million but can be as large as $50 million. OPIC's direct loans, reserved for foreign projects by small U.S. businesses, range from $500,000 to $6 million. OPIC will participate in up to 50 percent of the total cost for a new venture and up to 75 percent of an expansion. Call: (202) 408-9859, 1615 M Street, NW, Washington, DC 20527; Fax: (202) 336-8799.

The Enterprise Funds

The Enterprise Funds, established under the Support for Eastern European Democracy Act of 1989, promote private sector development in East Europe through equity investments, grants, insurance, guarantees and training. The Funds invest primarily with nationally-owned companies, joint ventures, and, less frequently, with subsidiaries of American companies with operations in Eastern Europe. The Funds will take an equity position in small and medium-sized ventures. Private partners must assume risk as well. The Funds prefer to have third-party investors in each project in order to leverage its capital.

1. Polish-American Enterprise Fund:
535 Madison Avenue, 33rd Floor, New York, NY 10022
Phone: (212) 339-8330; Fax: (212) 339-8359

2. Bulgarian-American Enterprise Fund:
   333 West Wacker Drive, Suite 2080, Chicago, IL 60603
   Phone: (312) 629-2500; Fax: (312) 629-2929

3. Hungarian-American Enterprise Fund:
   1620 Eye Street, NW, Suite 703, Washington, DC 20006
   Phone: (202) 467-5444; Fax: (202) 467-5469

4. Czech and Slovak-American Enterprise Fund:
   1620 Eye Street, NW, Suite 703, Washington, DC 20006
   Phone: (202) 467-5480; Fax: (202) 467-5466

5. Environmental Enterprises Assistance Fund: The Environmental Enterprises Assistance Fund, a non-profit venture established by the U.S. Agency for International Development and Winrock International, promotes the spread of commercially viable renewable energy and environmentally responsible technologies in developing countries. The Fund will invest in promising projects which, if they lead to profitable business, will return funds for further development. The Fund will provide technical and management assistance, loans, and equity and will act as a catalyst to overcome non-market barriers. The Fund will manage a $20 million portfolio.
   Contact: Environmental Enterprises Assistance Fund
   Suite 202, 1611 North Kent Street, Arlington, VA 22209
   Phone: (703) 522-5928; Fax: (703) 522-6450

INTERNATIONAL FINANCING

Multilateral Investment Guarantee Agency (MIGA)

The Multilateral Investment Guarantee Agency of the World Bank provides insurance against risks similar to that offered by OPIC. MIGA and OPIC will work together on a project. Czechoslovakia, Hungary and Poland are full members; Bulgaria and Romania have signed the MIGA convention but have yet to ratify it. MIGA can insure new investments between member countries where in project is in a developing country. New contributions to expand, modernize, or financially restructure an existing enterprise are eligible. MIGA will insure investments that privatize a state enterprise. Equity, loans, loan guarantees, licensing agreements, and technical assistance contracts may be insured.

Contact: Regional Manager/Europe, Multilateral Investment Guarantee Agency, 1818 H Street, NW, Washington, DC 20433; Phone: (202) 473-5419; Fax: (202) 477-9886.

International Finance Corporation (IFC)

The International Finance Corporation, an affiliate of the World Bank, is the world's largest source of debt and equity assistance to the private sector in developing countries. Bulgaria, Czechoslovakia, Hungary, Poland, Romania and the Balkans are eligible. The IFC invests in most types of product enterprises in its developing member countries. It usually limits its debt and equity participation to 25 percent of the project costs. The venture should cost at least $4 million in the smaller economies and $10 million in the larger ones. The IFC is able to arrange financial packages by coordinating financing from commercial banks, export credit agencies, and other institutions; facilitate syndicating large loans; and arrange debt for equity swaps.

Contact: Department of Investments, Director/Europe, International Finance Corporation, 1818 H Street, NW, Washington, DC 20433; Phone (202) 473-9634; Fax: (202) 334-8965.

INTERNATIONAL BANK FOR RECONSTRUCTION AND DEVELOPMENT (IBRD OR WORLD BANK)

In the fiscal year that ended June 30, 1991, the World Bank and its affiliate, the International Development Association, made loans and credits of $23 billion. Because the Bank typically finances one-third of any project, it assisted projects worth a total of $69 billion. Albania, Bulgaria, Czechoslovakia, Estonia, Hungary, Latvia, Lithuania, Poland, Romania, Russia and the CIS, and Yugoslavia are all members of the Bank.

Borrowing governments administer all loans and credits. Although borrowing countries have full responsibility for designing and implementing projects, the IBRD ensures that its procurement policies are followed throughout. Qualified companies in all member countries of the Bank are eligible to compete on all projects.

Companies wishing to bid on the IBRD's projects should maintain close liaison with the responsible ministry in the borrowing country. The numerous consulting services required by most projects are procured through procedures described in the Bank's Consultant Guidelines.

The World Bank offers a variety of publications, many free, to explain its procedures. The International Business Opportunities Service (IBOS), a subscription service, announces Bank projects.

Contact: World Bank Publications, P.O. Box 7247-8619, Philadelphia, PA 19170. Phone: (908) 225-2165; Fax: (908) 417-0482.

The Bank offers a business briefing every month. Contact: Business Relations Advisor, The World Bank, 1818 H Street, NW, Washington, DC 20433. Phone: (202) 473-1795; Fax: (202) 676-0578. The U.S. Department of Commerce has a liaison officer to deal with procurement issues at the World Bank: Janice Mazur, Room 20158, U.S. Department of Commerce, Washington, DC 20230. Phone: (202) 377-4322; Fax: (202) 377-3954.

European Bank for Reconstruction and Development (EBRD)

The European Bank for Reconstruction and Development (EBRD) was formally inaugurated on April 15, 1991, to assist Eastern and Central European countries make the transition toward open, market-oriented economies. The EBRD's primary purpose is to support the development of the private sector. Albania, Bulgaria, Czechoslovakia, Estonia, Hungary, the Independent States, Latvia, Lithuania, Poland, and Romania are all eligible for Bank projects. The EBRD will direct at least 60 percent of its funding to the private sector or to state enterprises that are shifting to private ownership. The remaining 40 percent will go to projects that improve the environment or the infrastructure. The bank will act both alone and together with other multilateral institutions and with commercial banks. The United States is the largest of the 41 shareholders in the Bank. American companies are eligible to compete for all projects the EBRD finances.

The Banks has a Merchant Banking Division (private sector activities) and a Development Banking Division. Inquiries should be directed to: European Bank for Reconstruction and Development, 122 Leaden Hall, London EC3V 4OL, United Kingdom, Phone: (44-071) 338-6424 or 6126; Fax: (44-071) 338-6100. The U.S. Department of Commerce Liaison with the European Bank is: Tom Kelsey, Phone: (44-71) 338-6569; Fax: (44-71) 338-6487.

PROJECT FINANCE

Developmental agencies such as USAID and TPD announce projects in the U.S. publication, Commerce Business Daily, P.O. Box 5999, Chicago, IL 60680; phone publications. Another source: Superintendent of Documents, U.S. Government Printing Office, Washington, DC 20402; Phone: (202) 783-3238; Fax: (202) 512-1356.

U.S. Agency for International Development (USAID)

USAID, under the American Business Initiative, will implement projects covering agribusiness, energy, the environment, housing and telecommunication to facilitate the transfer of U.S. commercial expertise and the flow of U.S. capital to Eastern Europe. The initiative is designed to increase the participation of small and medium-size American companies in this market.

Contact: Office of Development Resources, Bureau for Europe, U.S. Agency for International Development, Washington, DC 20523-0053; Phone: (202) 647-9734; Fax: (202) 647-6962.

USAID may provide funds to recipient governments, which will in turn issue tenders directly. Besides Commerce Business Daily, the publication, USAID Procurement Information Bulletin, available free, is a valuable source for USAID contracts: Office of Small and Disadvantaged Business, Minority Resource Center, U.S. Agency for International Development, Washington, DC 20523; Phone: (703) 875-1551; Fax: (703) 875-1862.

U.S. Trade and Development Program (TDP)

The U.S. Trade and Development Program (TDP), and independent agency of the U.S. Government, promotes economic development by funding feasibility studies, training programs, consultancies, and seminars in developing countries. In Eastern Europe, TDP identifies major projects in the public sector that offer large export or investment potential for American companies. TPD can fund American firms in project planning.

All proposed projects must have host government's development priority, U.S. export potential, and untied financing. It is up to host government agencies to request financing, either directly to TDP or through the local U.S. Embassy, but U.S. companies may bring these projects to the attention of TDP. TDP's programs are available in Bosnia-Hercegovina, Bulgaria, Croatia, Czechoslovakia, Estonia, Hungary, the Independent States, Latvia, Lithuania, Poland, Romania and Slovenia. Contact: U.S. Trade and Development Program, SA-16, Room 309, Washington, DC 20523; Phone: (703) 875-4357; Fax: (703) 875-4009.

U.S. Small Business Administration

The Small Business Administration offers loan guarantees to qualifying companies to assist them in entering international commerce or in expanding existing businesses. Some of these guarantees may cover working capital loans. Contact: Office of Business Loans, Small Business Administration, Sixth Floor, 409 Third Street, NW, Washington, DC 20024; Phone: (202) 205-6570; Fax: (202) 205-7519.

APPENDIX Y - THE EXPORT-IMPORT BANK

This appendix was made available by the Export-Import Bank of the United States, 811 Vermont Avenue, NW, Washington, DC 20571.

The Export-Import Bank of the United States (Eximbank) is and independent U.S. Government agency that helps finance the sale of U.S. goods and services to foreign buyers. Over the past 58 years, Eximbank has helped finance more than $220 billion in U.S. exports.

HOW DOES EXIMBANK WORK?

Eximbank supports export sales by making it easier for overseas buyers to purchase U.S. goods and services on credit. It does this in two ways: 1) Eximbank provides loans directly to foreign buyers; 2) It encourages U.S. suppliers or their banks to extend credit terms to foreign buyers by providing credit risk protection, and in some cases fixed-rate funding support. Eximbank also helps U.S. exporters obtain pre-export financing through its Working Capital Guarantee Program.

WHEN CAN EXIMBANK HELP?

Eximbank will support the sale of U.S. goods or services to a creditworthy foreign buyer when private financing cannot be arranged. To qualify for Eximbank support, the product or service must have at least 50% U.S. content, and cannot be military-related. All transactions also must be economically viable and must not affect the U.S. economy adversely. There must be a reasonable assurance of repayment.

While Eximbank is not a foreign aid or development agency, its programs often help U.S. exporters participate in development projects. Eximbank has also co-financed projects with the U.S. Agency for International Development (AID), the World Bank, and regional developments banks. If an overseas buyer is going to import goods or services, there may be an Eximbank program that can be used to finance purchases from the United States.

WHAT TYPES OF PROGRAMS DOES EXIMBANK OFFER?

1. Eximbank can guarantee 100% of the principal and interest on commercial loans to small and medium-sized companies that need funds to produce or market U.S. goods or services for export. Exporters may apply for Preliminary Commitments and then "shop" for lenders. This Working Capital Guarantee can be for a single export-related loan or a revolving line of credit. There is no minimum or maximum loan amount. The repayment period generally is 12 months, and it may be renewed.

2. Eximbank can provide loans—competitive, fixed-interest rate loans—directly to foreign buyers of U.S. capital equipment and services, or through an intermediary lender. Eximbank's intermediary loans enable commercial lenders or exporters to extend competitive, fixed-rate financing to their overseas buyers. Eximbank can also guarantee commercial loans (including convertible foreign currency loans) to overseas buyers of U.S. goods or services against both political and commercial risks of nonpayment.

Intermediary loans and guarantees cover the sale of capital items such as trucks and construction equipment, scientific apparatus, food processing machinery, medical equipment or projected-related services—including architectural, industrial design, and engineering services. Long-term loans and guarantees are available for major projects, large capital goods and/or project-related services.

Eximbank's loans and guarantees cover 85% of the contract price (100% of the financed portion) with the foreign buyer required to make a minimum cash downpayment of 15%. The fees charged by Eximbank for its programs are based on the nature of the product or service being exported, the creditworthiness of the foreign buyer and its country, as well as the comparable agencies of other exporting countries.

3. Eximbank can provide credit insurance policies for single or repetitive export sales or leases to foreign buyers through its agent, the FCIA Management Company, Inc. These policies provide protection against both the political and commercial risks that a foreign buyer might default on a credit obligation. A variety of policies are tailored for use by exporters and financial institutions. They generally cover 100% of the principal for political risks, and 90-95% for commercial risks, as well as a specified amount of interest. Exporters may obtain receivables financing more easily because the proceeds of the policy are assignable to banks. Short-term insurance policies are used to support the sale of consumer goods, raw materials and spare parts on terms of up to 180 days, and bulk agricultural commodities, consumer durables and capital goods on terms of up to 360 days. Capital goods may be insured for up to 5 years, depending upon the contract value, under medium-term policies.

WHERE TO APPLY FOR EXIMBANK PROGRAMS

There is wide access to Eximbank's loan and guarantee programs. Any responsible party—the foreign buyer, the U.S. exporter, a lending institution, or a firm representing either the buyer or the exporter—can apply directly to Eximbank for a Preliminary Commitment (PC) setting forth the terms and conditions under which Eximbank's loan/guarantee will support the export sale transaction. Similarly, export credit insurance is available through Eximbank's agent, the FCIA Management Company, Inc.

Further information about Eximbank programs may be obtained from a potential U.S. exporter, a potential foreign buyer's bank or its correspondent bank in the United States, the foreign buyer's government, U.S. embassies or consulates, FCIA Management Company, Inc. regional offices, state governments participating in the Eximbank City/State Program, or by contacting the Export-Import Bank directly, as follows:

Export-Import Bank of the U.S.
811 Vermont Avenue, NW
Washington, DC 20571-0999
Telephone: (202) 566-4490; Fax: (202) 566-7524
Cable: EXIMBANK; Telex: (TRT) 197681 EXIM UT

Export-Import Bank of the U.S.
West Coast Office
11000 Wilshire Blvd., Suite 9103
Los Angeles, CA 90024
Telephone: (310) 575-7425; Fax: (310) 575-7428

APPENDIX Z - THE OVERSEAS PRIVATE INVESTMENT CORPORATION

This appendix was made available by the Public Affairs Officer, OPIC, 1615 M Street, NW, Washington, DC 20527.

OVERVIEW

The Overseas Private Investment Corporation (OPIC) is a self-sustaining, U.S. government agency whose purpose is to promote economic growth in developing countries by encouraging U.S. private investment in those nations. By doing such, OPIC can help American companies remain competitive in the international marketplace.

OPIC assists U.S. investors through two principal programs: (1) financing investment projects through direct loans and/or loan guaranties, and (2) insuring investment projects against a broad range of political risks.

All of OPIC's guaranty and insurance obligations are backed by the full faith and credit of the United States of America, as well as by OPIC's own substantial financial reserves.

OPIC assistance is available for new business investments and expansion in more than 100 developing countries and areas around the world. However, OPIC will not provide assistance for any project that adversely affects the U.S. economy or domestic employment, is financially unsound, or does not promise significant benefits to the social and economic development of the host country.

The Overseas Private Investment Corporation was established by Congress in 1969, and began operations in 1971. Structured like a private corporation, it does not receive Congressional appropriations. Moreover, OPIC has recorded a positive net income for every year of operations, with reserves currently standing in excess of $1.3 billion.

OPIC's professional staff, recruited primarily from the private sector, is dedicated to responding quickly to investor needs and to developing creative business solutions for furthering U.S. private investment in the developing world.

PRIMARY PROGRAMS AND SERVICES

Financing: Medium- to long-term financing for overseas investment projects is available through loan guaranties and/or direct loans. OPIC's all-risk loan guaranties, issued to U.S. lending institutions, typically range from $2 million to $25 million, but can be as large as $50 million. OPIC's direct loans, reserved for overseas investment projects involving small and mid-sized companies, typically range from $500,000 to $6 million. In general, OPIC's finance commitments do not exceed 50 percent of the total project cost.

Political Risk Insurance: OPIC can insure U.S. investments overseas against the risks of political violence (war, revolution, insurrection, civil strife) to both assets and business income; expropriation; and inconvertibility of local currency. Specialized insurance coverages are also available for U.S. investors involved with certain contracting, exporting, licensing or leasing transactions to be undertaken in a developing country.

Investment Missions: OPIC traditionally conducts periodic investment missions to developing countries offering excellent investment opportunities for American businesses. Such missions are designed to introduce senior U.S. business executives to key business leaders, potential joint-venture partners and high-ranking government officials in the host country.

Opportunity Bank: This computer data system "matches" a U.S. investor's interest with specific overseas opportunities. American firms seeking joint-venture projects overseas submit a description of their company, the type of investment sought and the developing country or countries of interest. Upon request, the information is "matched" against similar information submitted by foreign businesses seeking American investors. Registration is free, with a modest fee charged for "match" requests. OPIC accepts registrations for this service without detailed check of the accuracy of reliability of the information submitted.

Investor Information Service: This information clearinghouse provides U.S. companies and individuals with basic economic, business and political information and data on 110 developing countries and 16 geographical regions. The information is packaged in country-specific and region-specific kits, available for a nominal fee.

APPENDIX AA - NEW SOURCES ON EASTERN EUROPE AND THE BALTICS

This appendix was made available by the Eastern Europe Business Information Center, U.S. Department of Commerce, Washington, DC 20230.

AmCham Today
Newsletter of the American Chamber of Commerce in Hungary.

American Chamber of Commerce
Budapest VI
Dozsa Gyorgy u. 84/a
Room 412
H-1068 Hungary
Tel: (011)(36)(1) 142-8752

AABS
Newsletter of the Association for the Advancement of Baltic Studies, Inc.

AABS Newsletter
111 Knob Hill Road
Hackettstown, NJ 07840
Tel: (908) 852-5258

BNA's EASTERN EUROPE REPORT
Bi-weekly newsletter covering changes in policies, laws, and regulations, as well as market deals in Eastern Europe.

Bureau of National Affairs, Inc.
1231 25th Street, N.W.
Washington, DC 20037
Tel: (202) 452-4200
Telex:285656 BNAI WSH
Customer Relations: 1-800-372-1033

Bulgarian Business News-168 Hours BBN
Weekly publication on recent issues concerning Bulgaria.

"168 Hours" Ltd
2A "Aleksandar Stamboloosky" blvd.
1000 Sofia
Bulgaria

The Bulgarian Watcher
Weekly newsletter on economic issues, political events, and business opportunities affecting Bulgaria.

The Bulgarian Watcher
1, Lazar Stanev
113 Sofia
Bulgaria
Tel: (3592) 706165
Fax: (3592) 720509

Business Club in Bulgaria
Magazine published 10 times per year covering developments in the Bulgarian economy.

Informa-Consult
55 Chapev Str.
1574 Sofia, Bulgaria
Tel: (3592) 718153
Fax: (3592) 704145

Business Czechoslovakia
Monthly newsletter on economic and political events and business opportunities.

The Rendon Group, Inc.
1802 T Street, NW
Washington, DC 20009
Tel: (202) 745-4900

Business Directory of Eastern Europe
Listing of over 3000 Eastern European companies still in business and interested in international trade.

American Business Press
P.O. Box 623
Lewiston, NY 14092-0623
USA

Business Eastern Europe
Weekly newsletter on market opportunities and business information.

Business International Ltd.
Customer Service Department
40 Duke Street
London WIA IDW
Tel: (011)(44)(71) 493-6711
Fax: (011)(44)(71) 491-2107

Business Tech International
A monthly news magazine designed to keep Romanian businessmen abreast of the latest business trends, future technology, and management techniques. Text is in Romanian.

Business Tech International
ASE World Enterprises
11 Asylum Street
Hartford, CT 06103
Tel: (203) 293-2714
Fax: (203) 549-7436

Business Tech Romania
A bi-monthly publication highlighting the business market in Romania.

Business Tech Romania
ASE World Enterprises
11 Asylum Street, Suite 614
Hartford, CT 06103
Tel: (203) 293-2714
Fax: (203) 549-7436

Central European
A monthly publication concentrating on East European business practices, joint ventures, secondary markets, accounting and banking updates and summaries of new laws and regulations.

International Information Services, Inc.
P.O. Box 3490
Silver Spring, MD 20918
Tel: (301) 555-2975
Fax: (301) 565-2973

Countertrade Outlook
Weekly intelligence on reciprocal international trade and unconventional trade finance.

Susanne O. Paine
DP Publications
Box 3141 Park Fairfax Station
Fairfax, VA 22039
Tel: (703) 425-1322
Fax: (703) 425-7188

Czechoslovak Business Update
Bi-monthly newsletter reporting on Czechoslovakia's economic transformation, foreign investment, and current business conditions.

Chris Pomery
Nerudova 37/8, 118 00 Praha 1
Czechoslovakia
Tel: (422) 53 25 66
Fax: (422) 53 25 66

Czechoslovakia Investment News
Bi-weekly newsletter on investment climate and opportunities in Czechoslovakia.

160 Woonsocket Hill Road
North Smithfield, RI 02895
Tel: (401) 769-4842

Czechoslovak Update
Monthly update on legal, economic, and political events in Czechoslovakia.

Central Europe Institute
1100 Connecticut Avenue, NW
Suite 500
Washington, DC 20036
Tel: (202) 296-9141
Fax: (202) 296-9153

Daily Report—Eastern Europe
Daily reports based on foreign radio and television broadcasts and news publications.

Foreign Broadcast Information
Service c/o National Technical
Information Service
5285 Port Royal Road Springfield,
VA 22161
Tel: (703) 487-4630
Fax: (703) 321-8547

Directory of Consultants and Contractors Active in Eastern Europe
Lists consultants and contractors in the areas of architectural and interior design, consulting engineers, consultants in urban and rural planning and construction contractors.

Projects Research, Inc.
P.O. Box 2558
Falls Church, VA 22042, USA
Tel: (703) 698-9330
Fax: (703) 698-9837

Directory of Foreign Trade Organizations in Eastern Europe
Over 450 pages of export/import companies and supporting organizations engaged in foreign trade as well as related companies in Eastern Europe.

Vance T. Petrunoff, Publisher
2 Townsend Street, #2-304
San Francisco, CA. 94107
Tel: (415) 979-0260
Fax: (415) 495-5843
Telex: 6404192419
MCI Mail: 419-2419

Directory of U.S. Companies Doing Business in the Soviet Union and Eastern Europe
Lists American companies with established business activities in the region and identifies essential sources of assistance for pursuing trade and investment there.

Wetherby International Company (IRS # 54-1533535)
P.O. Box 5393
Arlington, VA 22205
Tel: (703) 241-0586

Doing Business in Eastern Europe
Series of country specific guides to doing business in Eastern Europe. Updated monthly.

Vencor Publishing Corporation
1001 Third Avenue West
The Barnett Building, Suite 350
Bradenton, FL 34205
Tel: 813-746-5401
Fax: 813-747-5602

Doing Business with Eastern Europe
10 volume reference set on doing business with Eastern Europe with monthly updates. Single country additions are updated quarterly.

Business International Corp
215 Park Avenue South
New York, NY 10003
Tel: 212-460-0600
Fax: 212-995-8837

The East European Business Report
Biweekly Newsletter for the Global Manager

The East European Business Report
5842 Mayfield Road
Mayfield Heights, OH 44124-2934
Tel: (216) 449-0800
Fax: (216) 449-1506

East European Construction
Biweekly newsletter on construction and real estate development in Eastern Europe and the Soviet Union.

Finnexus International Co.
2140 Partridge Berry Road
Birmingham, AL 35244
Tel: (205) 987-2948
Fax: (205) 987-1812

Eastern European Business Directory
Over 8,000 Eastern European companies and organizations described.

Gale Research Inc.
P.O. Box 33477
Detroit, MI 48232-5477
Tel: 1-800-877-GALE
Fax: (313) 961-60g3

Eastern European Energy Report
Monthly publication focusing on energy and the environment in Eastern Europe.

Eastern European Marketing Group
171 Madison Avenue
Suite 401
New York, NY 10016

The East European Industrial Monitoring Service
Monthly publication on key industries, including a clipping service with translation and analysis of Eastern European publications.

Business International
214 Park Avenue South
New York, NY 10003
Tel: (212) 460-0600
Fax: (212) 985-8837

The East European Investment Monthly
Monthly publication focusing on acquisitions, joint ventures and greenfield investment in Eastern Europe and the former Soviet Union. This monthly also covers investment laws in each state.

The East European Monthly
667 Madison Avenue, 18th Floor
New York, NY 10021
Tel: (212) 355-2931
Fax: (212) 371-1920

East European Markets
Weekly publication on economic reform, legislation, business opportunities, statistics and EEC-Comecon news.

Financial Times Business Information, Ltd.
European Business Publications
P.O. Box 891
Darien, CT 06820
Tel: (203) 655-2701
Fax: (203) 655-8332

East-West Executive Guide
Monthly publication covering legal and business developments in Eastern Europe and the CIS

World Trade Executive Inc.
P.O. Box 761
Concord, MA 01742
Tel: 401-454-2000
Fax: 401-454-2005

East/West Technology Digest
Monthly digest of new technology from the Soviet Union and Eastern Europe

Welt Publishing Company
1413 K. Street, NW Suite 800
Washington, DC 20005
Tel: 202-371-0555

Eastern Europe: An Energy Report
Monthly publication on energy and environmental issues and markets in Eastern Europe.

Eastern European Marketing Group
Strategic Marketing Inc.
171 Madison Avenue, Suite 401
New York, NY 10016
Tel: (212) 213-8044

Eastern Europe Business Briefing
Bi-monthly newsletter on western interests in Eastern European markets.

Eurospan, Ltd.
Publications Group
P.O. Box 3045
Barrington, IL 60010-3045
Tel: (708) 202-0099
Fax: (708) 705-6575

Eastern Europe Business Bulletin
Published bi-monthly on business information and opportunities.

Eastern Europe Business Information Center
U.S. Department of Commerce, Room 7412
14th & Constitution Avenue, NW
Washington, D.C. 20230
Tel: (202) 377-2645
Fax: (202) 377-4473

Eastern Europe Demographics and Business Opportunities
Quarterly updated set of reports on business opportunities in Eastern Europe, including the telecommunications, computer and energy/ environmental sectors.

EDI-Enterprise Development International Inc.
5619 Bradley Blvd.
Bethesda, MD 20814
Tel: (301) 652-0141
Fax: (301) 652-0177

Eastern Europe Finance
Bi-weekly newsletter on sources of financing for Eastern Europe.

DP Publications Co.
Michael Morrison
Box 7188
Fairfax Station, VA 22039
Tel: (703) 425-1322
Fax: (703) 425-7911

Eastern Europe Finance Update
Quarterly update on banking and finance in Eastern Europe, including sources of finance and strategies for countertrade.

Business International
215 Park Avenue South
New York, NY 10003
Tel: (212) 460-0600
Fax: (212) 995-8837

Eastern Europe Newsletter
Bi-monthly publication on political and economic policy forecasts and assessments on Eastern Europe and the Soviet Union.

Eastern Europe Newsletter Ltd.
87 Duke Road
London W4 2BW, U.K.
Tel: (011)(44)(081) 995-3860
Fax: (011)(44)(081) 747-8802

Eastern Europe Report
Weekly newsletter on opportunities in trade and investment in Eastern Europe.

American Banker/Eastern Europe Report
P.O. Box 30240
Bethesda, MD 20824
Tel: (301) 654-5580
Fax: (301) 654-1678

Eastern Europe Strategist
A private publication on political, economic and social developments in Eastern Europe.

T.G. Cutler, Editor
31 Headley Road
Morristown, NJ 07960
Tel: (201) 538-5672

Eastern European & Soviet Telecom Report
Monthly newsletter highlighting country and company profiles in telecommunications, information processing, and broadcasting sectors.

International Technology Consultants
2940 28th Street, NW
Washington, DC 20008-3413
Tel: (202) 234-2138
Fax: (202) 483-7922

Economic Bulletin for Europe
Yearly summary of economic developments and policies in European economies. Subscriptions are not available, but complimentary copies may be obtained.

United Nations Information
Tel: (202) 289-8670

The Economist Intelligence Unit
Provides country reports on financial and economic developments in various Countries including those in Eastern Europe.

40 Duke Street
London, WIA IDW
Tel: 071-493-6711
Fax: 071-499-9767

Environment Watch- East Europe. Russia. & Eurasia
Monthly newsletter covering environmental trends, events, and opportunities in Eastern Europe and the Newly Independent States.

37 Broadway
Arlington, MA 02174-5539
Tel: (617) 648-8700
Fax: (617) 648-8707

Environment and Health in Central and Eastern Europe
Quarterly publication of the Program for Environment and Health at Management Sciences for Health in Boston, Massachusetts.

Program for Environment and Health
Management Sciences for Health
400 Centre Street
Newton, MA 02158
Tel: (617) 527-9202
Fax: (617) 965-2208

Europe Now
Informs U.S. businesses of the latest events in the EC and the opportunities being created by the formation of this new market.

Single Internal Market Information Service
Room 3036
U.S. Department of Commerce
14th Street and Constitution Avenue
Washington, D.C. 20230
Tel: (202) 377-5276
Fax: (202) 377-2155

European Business Directory - 1992
400 page resource guide to the EC and Eastern Europe - Updated annually

Faulkner and Gray, Inc.
106 Fulton Street
New York, NY 10038

European Update-Database on Eastern Europe
Detailed country -by-country reports on the evolving political situation

DRT Europe Services
27 Avenue des Arts
B-1040 Brussels, Belgium
Tel: (32.2)230-59-80
Fax: (32.2) 230-58-78

Europrompt
A management guide to trends in Central and East European IT industry.

PRIM information Technology Ltd.
Blaha Lujza ter 3. Budapest
1088, Hungary
Tel: (361)138-4505, (361)138-4300
Fax: (361) 138-4086

Eye on the East
Bi-weekly publication on events in Eastern Europe and the Soviet Union.

Pacific-Sierra Research Corporation
1401 Wilson Boulevard, Suite 1100
Arlington, VA 22209
Tel: (703) 527-4975

Finance East Europe
Published by the London Financial Times, this bi-weekly publication highlights developments in the areas of East European and (former) Soviet privatization, cross-border mergers, financing and banking

International Information Services, Inc.
P.O. Box 3490
Silva Spring, MD 20918
Tel: (301) 565-2975
Fax: (301) 565-2973

Financial Times
A daily newspaper published in London. Includes information on finance and investment in Eastern Europe and the CIS in addition to other news items of interest.

Financial Times
Number One Southwark Bridge
London, SEI 9HL
Tel: 071-873-3000
Fax: 071-407-5700
Tel: 800-547-8833 (For U.S. Subscriptions)

Foreign Broadcast Information Service
Daily or weekly publication featuring news accounts, commentaries, and Government statements from foreign broadcasts, press agency transmissions, newspapers, and periodicals published in the previous 48-72 hours.

National Technical Information Service (NIIS)
5285 Port Royal Road
Springfield, VA 22161
Tel: 703-487-4630

General Trade Index & Business Guide
Over 700 pages detailing the intricacies of the Polish legal and tax systems and listing potential Polish trade partners.

Business Foundation
Warszawa, ul. Wspolna 1/3
Tel: (0-22) 28-40-71 do 87 wew. 438
Fax: 28-05-49
Telex: 817088

Gazeta International
Weekly publication on Poland.

Gazeta Wyborcza
P.O. Box 348
New York, NY 10040
Tel: (212) 956-4040

Hungarian Rules of Law in Force
Full texts of Hungarian tax, trade and investment laws, published in English, German, and Hungarian.

International Information Services, Inc.
P.O. Box 3490
Silver Spring, MD 20918
Tel: (301) 565-2975
Fax: (301) 565-2973

Hungarian Trade Journal
Monthly periodical of the Hungarian Chamber of Commerce

Kultura
POB 149
H-1389, Budapest

Hungary International
A guide to companies and individuals interested in doing business in Hungary and other newly-liberated European countries.

Szablya Consultants Inc.
4416 - 134th Place, S.E.
Bellevue, WA 98006
Tel: (206) 643-1023
Fax: (206) 562-6381

Hungary Today
Monthly business, economic and political newsletter filled with timely information on the changing face of the country.

SURE Publishing Company
17 Morgan Court
Hurricane, WV 25526
Tel: (304) 757-8484

The Insider
Weekly newspaper focusing on economic, political and financial matters in Central Europe.

Dzielna lla/21
01-023 Warsaw, Poland
Tel: (48) (22) 38-33-44

International Trade Press
Catalog of various publications on international trade. Wide selection of publications available on East European trade, business, and investment.

International Trade Press
2 Townsend Street, #2-304
San Francisco, CA 94107
Tel: 800-359-6031
Fax: 415-495-4394

Invest in Hungary
Comprehensive Reportage and briefs on business environment in Hungary.

Business and Economy
Invest in Hungary
Subscription Department
H-1025 Budapest
Ruthen Utca 28
Hungary
Tel: (36/1)167-113, 180-4466
Fax: (36/1) 180-4466

PlanEcon Report
Weekly publication on economic situations in Eastern Europe.

PlanEcon
1111 14th Street, NW
Suite 801
Washington, DC 20005-5603
Tel: (202) 898-0471
Fax: (202) 898-0445

Plan-Econ Business Report
Bi-weekly newsletter that reports on business developments in the Soviet Union and Eastern Europe.

Marketing Department
Plan Europe Econ Ltd
Wimbledon Bridge House
1 Harthouse Road
Wimbledon SWI9 3RU, UK
Tel: 081-545 6206
Fax: 081-545 6233

Poland and Eastern Europe Report (PEER)
Monthly issue briefing on public and private sector political and economic issues and programs implemented in Eastern Europe.

W.M.C. Public Affairs Group
3000 K Street, NW, Penthouse 3-A
Washington, DC 20007
Tel: (202) 944-4920
Fax: (202) 944-4924

Polish Building Market
Monthly magazine detailing trends and prospects in Polish construction.

BIHC
00-099 Warsaw
ul. Senatorska 27
Poland
Tel: (48) (22) 27-24-49

Polish Business Voice
Monthly publication devoted to current political and economic changes in Poland and other Eastern European countries.

Krystyna Stachowiak, Managing Editor
Richard Cottrell & Associates International
655 15th Street, NW
Suite 300
Washington, DC 20005
Tel: (202) 639-4082
Fax: (202) 347-6109

Power Plays
A newsletter highlighting developments, products and news of interest to companies interested in power generation, pollution control and environmental service markets of Eastern Europe and the Soviet Union.

PenWell Publishing Powerplays
P.O. Box 1260
Tulsa, OK 74101
Contact person: Joseph Bessette
Tel: (918) 835-3161
Fax: (918) 831-9497

Privatization Newsletter of Czechoslovakia
Irregular, but generally monthly newsletter highlighting recent developments of the Czech and Slovak Federation's privatization program.

The Center for International Private Enterprise
1615 H St., NW
Washington, DC 20062
Tel: (202) 463-5901
Fax: (202) 887-3447

Resources CSFR.
Information directory for Czechoslovakia, updated every three months.

Resources CSFR
Lazarska 3
110 00 Praha 1
Czechoslovakia
Tel: (42/2) 201-325
Fax: (42/2) 201-812

Review of Commerce
A newsletter providing information on opportunities in various sectors in Poland, including information on joint ventures.

The Review of Commerce
60-529 Poznan
ul. Dabrowskiego 77
Tel: 426-26
Fax: 424-55
Telex: 041-3214

Romania Economic Newsletter
Report which reports on and analyzes economic developments in Romania.

COSMOS, Inc.
7201 Wisconsin Avenue
Suite 500
Bethesda, MD 20814, USA
Tel: (301) 229-5875
Fax: (301) 229-5876
Telex: 440184 DEVR. UI.

Slovakia
Information bulletin on legislation, investment, foreign capital, privatization, public tenders, banking and traditions. Published by Slovak National Agency for Foreign Investment and Development

Slovakia
Slovak national Agency for Foreign Investment and Development
Manesovo Namestie 2
851 01 Bratislava
IC O 17329868
Tel: (427) 224 575
Fax: (427) 225 266

Soviet Business and Trade with News From Eastern Europe
Twice monthly review of business and trade

Welt Publishing Company
1413 K Street, NW Suite 800
Washington, DC 20005
Tel: 202-371-0005

Strategic Business Report: East Europe/USSR (formerly "Glasnost")
Bi-weekly publication on recent developments in Soviet and Eastern European business, economics, and related legislation.

Continental Information North America
1011 Western Ave., Suite 900
P.O. Box 1076
Seattle, WA 98111-1076
Tel: (206) 382-9999
FAX: (206) 464-0125

Strategic Business Report Bulgaria.
Monthly publication focusing on the latest developments in Bulgarian commerce and finance.

Continental Information North America
1011 Western Ave., Suite 900
P.O. Box 1076
Seattle, WA 98111-1076
Tel: (206) 382-9999
FAX: (206) 464-0125

Tracking Eastern Europe.
Weekly newsletter on business and investment opportunities in Eastern Europe.

A.M.F. International Consultants
812 North Wood Avenue
Linden, NJ 07036
Tel: (201) 486-3534
Fax: (201) 486-4084

Transition
The newsletter of economic reform in the emerging democracies.

Richard Hirschler
Room N-6027
The World Bank
1818 H Street, N.W.
Washington, D.C. 20433
Tel: (202) 473-6982
Fax: (202) 676-0439

U.S.-Czechoslovakia Business News.
Quarterly newsletter on business activities and opportunities in Czechoslovakia.

ARA-PRAHA/NEW YORK
116 West 72 Street -16 E
New York, NY 10023
Tel: (212) 787-7644
Fax: (212) 787-3164

U.S. Department of State Dispatch
Weekly magazine provides comprehensive information on U.S. foreign policy, including fact sheets, country summaries, and updates on public and private sector assistance to Central and Eastern Europe.

National Technical Information Service (NIIS)
5285 Port Royal Road
Springfield, VA 22161
Tel: 703-487-4630

U.S. Poland Connection
Published by the U.S. Poland Chamber of Commerce promoting trade and investment between the United States and Poland.

U.S. Poland Connection
U.S. Poland Chamber of Commerce
321 North Clark Street
Suite 550
Chicago, IL 60610-4715
Tel: (312) 467-0614
Fax: (312) 467-0615

The Warsaw Voice

Weekly publication in English on the Polish economy, commercial laws, and opportunities.

The Warsaw Voice
413B Logan Boulevard
Lakemont
Altoona, PA 16602
Tel: (800) 488-2939

Wiadomosci Handlowe

Quarterly magazine highlighting American business activity in Poland.

U.S. Trade Development Center
ul.Wijeska 20 00-490 Warsaw, Poland
Tel: (48) (22) 21-45-15

The World Bank Watch

Weekly publication on World Bank activities in Eastern Europe and around the globe.

The World Bank Watch
P.O. Box 30240
Bethesda, M.D. 20824
Tel: (800) 733-4371; (301) 654-5

APPENDIX BB - ENTRY AND VISA REQUIREMENTS

POLAND

Entry Requirements: US citizens require
- Valid passport is required.
  - Effective April 15, 1991, visas are no longer required for a stay of up to 90 days. Extensions of stay may be applied for at the Voivodship Office of the Department of Internal Affairs. Assistance can be arranged for through an ORBIS office or hotel receptionist.
- Visa
- Proof of funds for expenses
  - confirmed, prepaid hotel voucher OR
  - currency exchange @ $15 per day, per person OR
  - ORBIS vouchers
    - contact Orbis Travel Bureau in New York, 500 Fifth Avenue, New York, NY 10016, 212-391-0844 or 730-1978.

Transit without visa is permitted if: arriving by plane, continuing to a third country by plane on the same day, staying in the airport transit area.

BUSINESS:
- 1 application form (for 1, 3, 4 or multiple entries)
  - photocopies are not acceptable
  - 2 forms required for double entry
- 2 photographs; 4 photographs required for double entry
- Valid passport
- Letter from employer stating purpose of travel and financial guarantee of expenses and return of onward transportation
- Invitation letter, if available
- Ministry authorization, if multiple entry ($20 telex fee)
- Visa fee
  - $23 1 entry
  - $43 2 entries
  - $63 3 entries
  - $83 4 entries
  - $120 multiple entries

TOURIST:
- 1 application form (for 1, 3, 4 or multiple entries)
  - photocopies are not acceptable
- 2 photographs; 4 photographs for double entry
- Valid passport
- Notarized letter of consent for persons less than 18 years old, if traveling unaccompanied by parents and letter must be signed by unaccompanying parent/s.

EMBASSY: 2640 16th Street, NW, Washington, DC 20009

TELEPHONE: 202/234-3800, 3801, 3802

HUNGARY

Entry Requirements:
US citizens require
Valid passport
International driver's license, if arriving by car
No vaccinations

BUSINESS/TOURIST/TRANSIT
Effective April 1, 1991, visas are no longer required for a stay of up to 90 days. Extensions of stay may be applied for in Hungary at the local police station.

EMBASSY: 3910 Shoemaker Street, NW, Washington, DC 20008.

TELEPHONE: 202/362-6730

CZECH AND SLOVAK REPUBLIC

Entry Requirements:
NOTE: Effective May 15, 1990, visas are no longer required for US citizens. Travelers may enter the Czech and Slovak Republic upon presentation of the following items:

Valid passport is required. Permission to stay up to 30 days is granted on arrival to tourists and businesspersons.
$25 fee
Vaccinations are not required for entry from any country.
Extensions of stay may be applied for through the local Passport and Visa authorities for up to 6 months.

NON-US CITIZENS which require visa must submit the following:

* 1 application form, per entry
* 2 photographs, per entry
* Valid passport
* $ visa fee (fluctuates)
* Letter of intent from employer, if on business

PROCESSING TIME: usually 1-2 days

VALIDITY: Valid for 1-2 entries within 5 months for a stay of up to 30 days, as requested on the application form. Extensions of stay may be applied for in the Czech and Slovak Republic.

EMBASSY: 3900 Linnean Ave., NW, Washington, DC 20008.

TELEPHONE: 202/363-6315 and 6316.